THE AUDITORY SYSTEM IN SLEEP

THE AUDITORY SYSTEM IN SLEEP

RICARDO A. VELLUTI

Neurofisiología
Departamento de Fisiología
Facultad de Medicina
Universidad de la República
Montevideo
Uruguay

AMSTERDAM • BOSTON • HEIDELBERG • LONDON • NEW YORK • OXFORD
PARIS • SAN DIEGO • SAN FRANCISCO • SINGAPORE • SYDNEY • TOKYO
Academic Press is an imprint of Elsevier

Academic Press is an imprint of Elsevier
84 Theobald's Road, London WC1X 8RR, UK
30 Corporate Drive, Suite 400, Burlington, MA 01803, USA
525 B Street, Suite 1900, San Diego, CA 92101-4495, USA

First edition 2008

Notice
No responsibility is assumed by the publisher for any injury and/or damage to persons
or property as a matter of products liability, negligence or otherwise, or from any use
or operation of any methods, products, instructions or ideas contained in the material
herein. Because of rapid advances in the medical sciences, in particular, independent
verification of diagnoses and drug dosages should be made.

British Library Cataloguing in Publication Data
A catalogue record for this book is available from the British Library

Library of Congress Cataloging-in-Publication Data
A catalog record for this book is available from the Library of Congress

ISBN: 978-0-12-373890-5

For information on all Academic Press publications
visit our web site at books.elsevier.com

Typeset by Charon Tec Ltd (A Macmillan Company), Chennai, India
www.charontec.com

Printed and bound in the United Kingdom

Transferred to Digital Print 2011

Working together to grow
libraries in developing countries

www.elsevier.com | www.bookaid.org | www.sabre.org

ELSEVIER   BOOK AID
             International   Sabre Foundation

*This book is dedicated to my closest family:
my wife Marisa, my daughters Rosina and Francesca,
my sons Alejandro and Federico, as well as my two
grandchildren, Valentín and Julieta.*

Contents

Foreword

Professor Ricardo Velluti is a scientist whose scholarship is based on vast scientific culture and successful experimental research. Early in his scientific career he took an interest in sleep and auditory physiology. This combination of topics brought about a deep insight into the mechanisms of information processing during the different behavioral states of the wake–sleep ultradian cycle.

There is a particular aspect of auditory processing which deserves attention with respect to sleep. Namely, the fact that, among the many sensory channels informing the brain, the auditory one is open not only during wakefulness but also during sleep. The importance of this auditory peculiarity may be appreciated by considering the fact that the auditory input is the most important information underlying the telereceptive interaction with the environment during sleep. From a general evolutionary viewpoint, it is obvious that this remarkable property of the auditory system entails survival relevance in many animal species. On this basis, the research carried out in Velluti's laboratory concerned the influence of acoustic stimulation on sleep behavior and the mechanisms of neuronal processing of acoustic information along the complex auditory pathways of the central nervous system during the ultradian wake–sleep cycle.

Leaving the appreciation of this book's content to the reader, I wish to add a personal note regarding Ricardo Velluti's work.

Our long-lasting collaboration allowed me to witness his full dedication to science and continuous striving for further progress in his field of study. His experimental cleverness and cultural scholarship put him high in writing a novel approach to the modulation and processing of auditory inputs in different behavioral states. In conclusion, I believe that his attractive and up-to-date book will be very useful to a large audience of readers including not only basic and clinical scientists but also students eager to expand their knowledge of new scientific developments in sleep and auditory physiology.

Professor Pier Luigi Parmeggiani
Bologna, Italia

June 14, 2007

Acknowledgements

It is clear to me how indebted I am to many people who joined me along my way through Neuroscience. To Professor Elio García-Austt, in whose laboratory I have worked since 1960, first as an MD student and soon after as a member of the staff, goes my most profound gratitude for the innumerable kindnesses he bestowed upon me for many years. He was my mentor and the person who really taught me how to think. We were not only engaged in many projects in Montevideo and afterward in Madrid, but we were also very good friends sharing most episodes of our lives, the good and the bad ones. I will always remember our sailing boat trips through the dangerous Río de la Plata.

Raúl Hernández-Peón, the brilliant Mexican Neurophysiologist, was also part of my endeavor when I worked with him at the Tlalpan laboratory in México DF. Our paths crossed again some time later at New Haven and that was the last occasion I saw him as he, still young, suddenly died soon after. He started me on sleep studies, a subject that had been haunting me since I read in a newspaper about Hess' research on that field, in the early 1950s.

During 1966–1967, after obtaining an NIH post-doctoral fellowship, I was accepted in Robert Galambos' laboratory at Yale University. This was an especially significant stay because of his scientific approach to Neuroscience at that time, and also because we have had personal exchanges on many subjects. We met again after a long while at San Diego for lunch and catching up. He asked me then what I thought, so many years after, about the auditory efferent system. What I told him was, in part, what is now written in this book.

Carmine Clemente invited me during a Latin American Physiology Congress in México City to collaborate at the Anatomy Department, Brain Research Institute at UCLA, where I arrived through an IBRO/Unesco fellowship. The period of 1970–1971 was a great time to be at Los Angeles. Sleep was then a *prima donna* of Neuroscience, which was very profitable for me. Talking with Clemente and other salient researchers who worked at UCLA at that time very much instructed and inspired me. We got together again some time ago at his office still in UCLA, where we remembered those times spent in close collaboration.

Secondly, I am indebted to my collaborators at the Neurophysiology laboratory of the Facultad de Medicina, Universidad de la República. All of them are cited in this little book because every one that came to my laboratory was a potential researcher.

I will just list them in alphabetical order: Claudia Bentancor, MD; Marcel Bouvier, MD; Luis Crispino, MD; Daniel G. Drexler, MD; Atilio Falconi, PhD; María I. Ferreira, MD; Juan P. Gambini, MD, Magister; Natalia Goldstein-Daruech, MD; Jack Hadjez, MD; Alejandra Inderkun, MD; Tamara Liberman, Magister; Carolina P. López, MD; Gabriela Morales-Cobas, MD; Lucía Pérez-Perera, Magister; Alberto Rodríguez Alvez, Lic; Zulma Rodríguez-Servetti, Lic; Pablo Torterolo, MD, Magister; Jack Yamuy, MD.

The two more relevant coworkers and close associates were José L. Peña, MD., PhD. and Marisa Pedemonte, MD., PhD. Both have contributed in an important fashion to the laboratory and in my personal development because of the many discussions, comments, and readings that came up every day. Dr. Peña is now at Yeshiva University, New York, and Dr. Pedemonte is a

Professor of Physiology at CLAEH University in Punta del Este, Uruguay.

Marisa, my wife and the mother of our two wonderful daughters, Francesca and Rosina, is also my immense companion in the adventure that is Science, and thus, my everything.

Two scientist collaborators and great friends for more than 20 years come now to my mind: Pier Luigi Parmeggiani and Peter M. Narins. Both were most important in my scientific and everyday life as well. They generously gave me their open friendship, something that I am continuously enjoying because we are still exchanging scientific and, most significantly, non-scientific ideas, and sharing visits, music concerts, congress meetings, and many other activities, in Montevideo, Los Angeles, or Bologna.

Montevideo,
August 2007

Introduction

The brain is a very complex information processing device, whatever the information it may be working on. The sensory input represents the whole fan of information the central nervous system (CNS) receives, to elicit, after complex processing, its output responses, e.g., motor, endocrine, neurovegetative, behavioral activities, or changes in the CNS capacities such as memory, learning, and so on. The information coming from the outer and the inner worlds during life is a meaningful influence on the brain phenotypic development and, in my particular topic, on sleep organization. An important purpose of the brain evolution is to allow the organism to properly interact with the environments, the external and the internal one (the body). In early developmental stages, from phylogenetic and ontogenetic viewpoints, the sensory information constitutes a relevant drive that controls the brain function and the general physiology in many ways. The development of each brain is genetically conditioned although a germane component is the continuous information coming in through the senses from both the worlds, a phenomenon that continues throughout life, i.e., an endless process.

Since the sensory information in general is continuously reaching the CNS, its processing will be differentiated according to the current physiological state of the brain, during: (a) wakefulness, (b) sleep stages I–IV, and (c) paradoxical sleep.

An important point that should be added is that the brain itself can condition its own sensory input by controlling all receptors and nuclei through the sensory efferent systems, which are present in every incoming pathway. Thus, by using this feedback possibility, the complex processing circuit may be completed through a functional "closed-loop" system.

The natural light–dark sequence, phylogenetically archaic information, through the light receptor and its processing system, profoundly influences the sleep–wakefulness cycle. The circadian rhythm of melatonin, produced in most organisms from algae to mammals, is generated in the latter by a central pacemaker located in the suprachiasmatic nuclei of the hypothalamus, largely synchronized by cues from the light–dark cycle. Thus, since the beginning of life, the brain and sensory systems complexity are in constant and mutual enrichment from both anatomical and functional perspectives. The auditory, olfactory, vestibular, and somesthetic system also developed introducing more sensory data which progressively shaped a brain that began to reach its completion, leading to a dynamic end: the genetically established sleep–waking cycle features.

Early in the 20th century, the concept of sleep as the result of a blockage of the auditory inflow was introduced while, later on, it was proposed that the extensive deafferentation of ascending sensory impulses to the isolated brain resulted in sleep. In any case, I am now putting forward that sensory information about the environment and the body continuously modulates the CNS activity during the sleep/wakefulness cycle.

Many inputs participate modulating the waking life and influencing sleep, which led to postulate that wakefulness is supported by the sensory systems and that a lack or decreased level in their activity would lead to sleep, constituting the "passive" sleep theory. In addition, interactions between sleep and sensory input in general were reported, and a surgical *quasi*-total deafferentation revealed significant changes in the characteristics of sleep and wakefulness.

Although profoundly modified, the processing of sensory information is still present during sleep. While all sensory systems show some influence on sleep, they are also reciprocally modulated by the sleep or waking state of the brain. Thus, the incoming sensory information may alter sleep and waking physiology and, conversely, the sleeping brain imposes rules on information processing. Although we do not yet completely understand how the brain processes sensory information, it is currently accepted that neuronal networks/cell assemblies can change depending on the information they receive throughout life.

Early in human life, auditory data could be recorded. A never published experimental approach carried out during 1966 at the Perinatology Centre of the University Hospital in Montevideo, showed the presence of auditory evoked potentials at very early stages in human life. During a human mature fetus delivery, when the uterus dilation was about 5 cm, two regular electroencephalography sterilized electrodes were attached to the fetus scalp by the obstetrician in charge. Through a small and shielded loudspeaker, clicks were delivered over the mother's womb surface while recording a clear-cut averaged fetus auditory evoked potential using a CAT computer.

Why the auditory system?

Several reasons support the notion of relating sleep to auditory physiology.

First, hearing is the only telereceptive sensory modality relatively open during sleep in micro-osmatic animals, acting as a continuous monitor of the environment, e.g., predator detection or a baby's cry during the night that awakes the parents.

Second, the auditory system has a conspicuous efferent component that makes it unique, featuring a complex anatomy located in parallel to the classic ascending pathway, and functioning as an input controller particularly through the action of its most peripheral sites, the olivo-cochlear system.

Third, the auditory stimuli can affect human sleep, e.g., a noisy night will reduce the total sleep time and will be followed by sleep normalization once the noise is reduced.

Fourth, the total lack of auditory input after bilateral surgical lesion of the cochleae alters the sleep architecture of guinea pigs and hamsters by increasing the total sleep time.

Fifth, the link between auditory memory traces and sleep is also expressed by the presence of auditory images in 65% of recalled dreams.

Sixth, the local blood flow is significantly increased in auditory *loci*, such as the auditory cortex, medial geniculate nucleus, inferior colliculus, superior olive, and particularly, the cochlear nucleus.

Seventh, imaging using functional magnetic resonance (fMRI) showed that the presentation of auditory stimuli in slow wave sleep elicited a significant bilateral activation in the auditory, parietal, and prefrontal cortices as well as thalamus and cingulate.

Eighth, human and animal auditory evoked potentials as well as evoked magnetoencephalographic activity change waveforms and amplitudes on passing to sleep. Another electrophysiological approach such as auditory single unit recording during sleep, also exhibited major shifts on passing to sleep (Chapter 5).

Including the auditory information processing during sleep is a step forward toward one of the probably main functions taking place in sleep, e.g., memory storage and probably some kind of sleep learning. These high functions perhaps performed during sleep are part of the scenery of this still enigmatic behavioral condition.

This book begins with Chapters 1 and 2 aiming at providing information about the two main approaches and trying to expose basic ideas on both the auditory system and the sleep behavioral state. It is not usual to analyze how the sensory information works during sleep because of many non-scientific prejudices.

The brain is just one and everything that takes places within it must have influences on several other systems, e.g., the sleep and auditory functional relationships.

Chapter 3 is engaged in notes about information processing and neuronal networks intending to introduce us in a general and mainly theoretical approach to our problem.

Chapter 4 is a general view of the many experimental ways and results to analyze the sleep–auditory system relationship. Evoked potentials, local-field and far-field human recorded potentials, magnetoencephalography as well as imaging are different points of interest exposed.

In Chapter 5, I am trying to approach the problem of the auditory processing by showing single units, their firings, their firing patterns, and their relationship with a well-known brain rhythm, the hippocampal theta rhythm. The possible auditory influences on sleep are part of Chapter 6. The auditory system constitutes an important way of incoming information that is not close during sleep, and that is why it can be altered by sound.

Finally, some conclusions will be made:

- from an auditory viewpoint;
- from a sleep point of view.

1

Brief analysis of the organization of the auditory system and its physiological basis

The auditory system with its associated anatomical and functional complexity subserves diverse processes such as discrimination of sound frequencies and intensities, sound source location in space, auditory learning, development of human language, auditory "images" in dreams, music, development of birds songs, i.e., communication in general. In this chapter a review of the known auditory ascending and descending systems is presented along with some new or not well-known approaches included.

The afferent ascending system

This complex system begins at the receptors in the cochlea followed by a wide upward expansion throughout the different nuclei, reticular formation, cerebellum, and connections to the primary and secondary cortices. It is composed of several neuronal groups with profuse communication from the cochlea to the cortex.

A non-classical ascending pathway branches off from classical IC and reaches the medial geniculate nucleus, medial and dorsal regions, to project to cortical regions (Moller and Rollins, 2002).

A diagram of the most important pathways and synaptic stations of the afferent auditory system is shown in Figure 1.1. The first-order auditory neurones, with cell bodies located in Corti's ganglion, send their axons centrally to form the auditory nerve, part of the VIIIth cranial pair. These nerve fibers synapse with the secondary neurones located centrally in different cochlear nucleus (CN) *loci*, in the medulla–pontine region. Let us bear in mind that 95% of the fibers which form the auditory nerve originate at the inner hair cells. The outer hair cells are innervated by only 5%, non-myelinated afferent thin fibers.

The auditory pathway has been described by using different methods of study throughout history: cell damage and degeneration, intracellular dyeing with tracers, deoxyglucose, and so on, and also by electrophysiological recording methods. By placing recording electrodes in various central nuclei, bioelectrical responses – changes in the membrane potentials – can be obtained of the auditory neurones which form the basis of evoked potentials measurable with gross electrode. Evoked potentials, recorded in cats, shown in Figure 1.3(A), are examples of the averaged responses to brief (click) sound stimuli. The differences between their shapes and, mainly, their latencies carefully reproduce the anatomical pathway, due to the fact that activity evoked by a stimulus first activate the receptors, followed by the auditory nerve fibers, and subsequently the central nervous system (CNS), orderly ascending from nucleus to nucleus.

Auditory nerve evoked activity

Beginning with the incoming sound, Table 1.1 exhibits the main mechanobioelectrical steps toward evoking an auditory nerve action potential. The auditory nerve compound action potential (cAP) can be recorded from an electrode placed at the round window. A cAP averaged is depicted in Figure 1.2 (left)

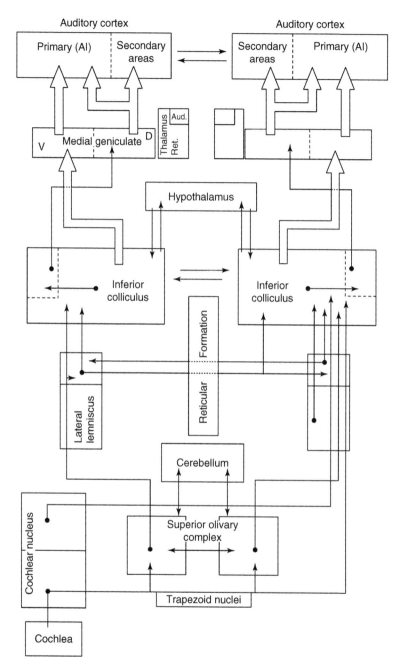

FIGURE 1.1 General diagram of the ascending auditory pathways. These pathways are embedded in the brain, an important concept since it reflects the multiple communication pathways that may affect incoming auditory information.

TABLE 1.1   Events toward the generation of an auditory nerve action potential

Sound waves move the tympanic membrane
↓
The tympanic membrane moves the middle-ear ossicles
↓
The ossicles move the oval window membrane
↓
The oval window movements produce motion of the cochlear fluids and basilar membrane
↓
The cochlear fluids and basilar membrane motion bend the inner hair cells' cilia
↓
The ciliary movements determine the excitation of the hair cells
↓
Finally, action potentials are generated at the auditory nerve fibers

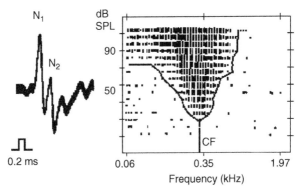

FIGURE 1.2   Guinea pig auditory nerve-averaged cAP recorded from the round window during wakefulness (left) and the tuning curve of one auditory nerve fiber recorded with a microelectrode (right). CF, characteristic frequency.

with the two classical negative waves N_1 and N_2, in response to clicks, i.e., a stimulus that synchronizes the discharge of many nerve fibers. It reveals the activity of a group of single fibers and its synchronized discharges. The N_1 amplitude is a function of the stimulus intensity as well as the number of synchronized fibers.

An auditory nerve single-fiber recording is shown in Figure 1.2 (right). A microelectrode may record the single-fiber activity when a stimulus of sufficient intensity is delivered. Its response can be characterized by a point of maximum sensitivity, i.e., the response at the stimulus frequency with the lowest intensity, the characteristic frequency. The complete tuning curve includes the range of a fiber's threshold discharges over a range of stimulation frequencies.

Evoked activity

The different nuclei in the auditory pathway exhibit evoked potentials with characteristic waveforms and latencies according to their position in the brain. They are the first approach to how the brain *loci* may process information. The signal amplitude and the anatomical location of the recoding macroelectrode give us a first indication about the processing. Figure 1.3(A) shows a series of local field recorded from a cat with electrodes placed in the auditory nuclei and cortex during wakefulness (W). The lower trace corresponds to the CN activation with the shorter latency, while the upper trace shows the primary auditory cortical response exhibiting the longer latency. The two evoked potentials recorded from the right and left lateral olivary nuclei present a different waveform depending on the ear being stimulated. This is the lowest auditory nucleus in which the system may be acquainted with the side from which the sound is coming, indicating the initiation of the auditory binaural signals analysis.

Evoked resistance shifts associated with the evoked potentials in anesthetized and awake cats were also recorded in parallel demonstrating their close functional relationship. Time locked to the evoked potential is the evoked resistance shift consisting of a brief resistance drop followed by a more prolonged rise recorded at all auditory nuclei but exhibiting

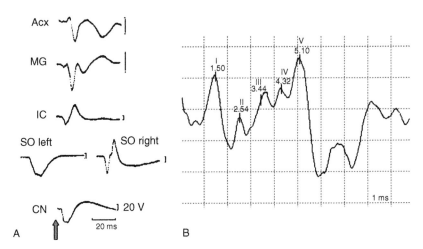

FIGURE 1.3 (A) Auditory local-field evoked potentials from a cat and (B) human far-field evoked responses. An increasing latency is observed as the recording electrode is located higher in the pathway. The arrow shows the time of the auditory stimulus (click). The human far-field recording shows from I (auditory nerve) to V (~inferior colliculus) the normal latencies in ms and the waves that roughly reflect the auditory nuclei in the brainstem. Acx, auditory cortex; MG, medial geniculate; IC, inferior colliculus; SO, superior olive; CN, cochlear nucleus.

a different wave shape and timing relative to the evoked potentials (Galambos and Velluti, 1968; Velluti and Galambos, 1970).

The far-field evoked potentials, mainly recorded in humans with scalp electrodes relative to mastoid electrodes, reveal a set of waves that correspond to an invading wave of negativity when sound stimuli are presented, i.e., are the responses of the negative activity throughout the brainstem beginning with the first wave that corresponds to the auditory nerve (I) activation. Figure 1.3(B) shows the short latency far-field potentials (brainstem waves I to V). Recordings with a different time scale (50ms) reveal the middle latency waves corresponding to the thalamic and cortical responses. A 500-ms time scale shows the full response including the late cortical potentials (not shown).

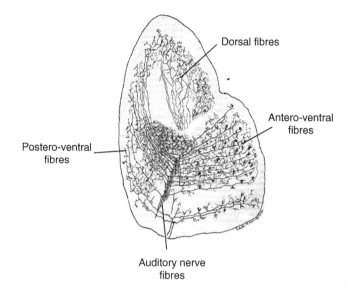

Dorsal fibres

Antero-ventral fibres

Postero-ventral fibres

Auditory nerve fibres

FIGURE 1.4   CN neuroanatomical approach (Lorente de Nó, 1981).

Cochlear nucleus

After entering the medulla, the auditory nerve fibres divide into two branches called anterior and posterior. The classical division of the CN is shown in Figure 1.4 in which the anterior branches innervate the ventral CN while the posterior ones end in the posterior and dorsal CN (Lorente de Nó, 1981). The tonotopic organization coming from the cochlea is totally maintained in the CN as well as all throughout the auditory pathway.

The neurones located in each sub-nucleus exhibit diverse bioelectrical properties producing different excitatory neurotransmitters such as glutamate and aspartate, and gamma aminobutiric acid (GABA) and glycine as inhibitory ones (Wenthold and Martin, 1984; Peyret et al., 1987).

The electrophysiological classification of the CN unitary firing is based on discharge patterns (Pfeiffer, 1966). Figure 1.5 shows examples of them: (A) primary-like, (B) onset,

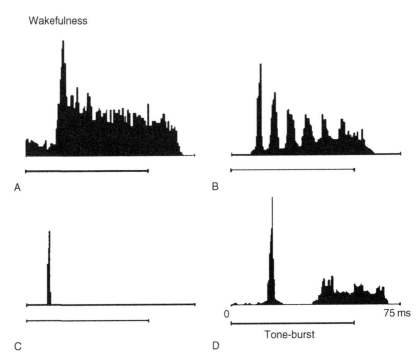

FIGURE 1.5   CN neurones post-stimulus time histograms (PSTH) classification: (A) primary-like, similar to the auditory nerve fibers discharge; (B) chopper response; (C) on response; (D) pauser, response to sound with a pause.

(C) chopper, (D) pauser. The firing pattern is not fixed, for each neuronal type, but rather is a function of stimulus intensity and the animal behavioral state. Furthermore, the CN exhibits processes of excitation–inhibition of diverse origin, some intrinsic from the CN itself, while others may arrive from higher centers such as the inferior colliculus (IC) and even via direct projections from the auditory cortex (Schofield and Coomes, 2005).

The CN unitary neuronal activity, as well as at any other CNS region, is a representation of one or more network's activity related to an auditory unit whose firing is always changing depending on several possibilities, e.g., stimulation characteristics, momentary behavior, etc.

Binaural fusion and localization

Listening to the same sound delivered to both ears, humans perceive a single phantom source between the ears or sometimes straight ahead in space. When sound to one ear is delayed, the source appears to shift to the other side. The clarity and compactness of perceived sources depend on the degree of similarity between the two signals. Addition of random noises to the signals can systematically vary the degree of binaural similarity. Such experiments showed that humans could perceive phantom sources even when the correlation of signals between the ears was as low as 0.3 (Jeffress et al., 1948). Humans are not the only species that is so good at detecting binaural correlation in signals. Barn owls are known for their ability to locate prey by ear. Owls rapidly turn their heads toward sound sources, even when signals are delivered by earphones. This simple observation means that owls also perceive a single phantom source from binaural signals. Thus, tests similar to those used for humans could be applied to owls. The results were very similar to those obtained in humans. Owls could localize signals with binaural correlations as low as 0.3. The standard deviation of the mean localization errors remained almost constant from correlation 1 to correlation 0.3, below which the standard deviation increased sharply (Saberi et al., 1998).

The study with owls further advanced our knowledge about the neural mechanisms underlying binaural fusion and localization. The barn owl's brainstem auditory system consists of two pathways, one for processing the interaural time difference (ITD) and the other for the interaural intensity difference (IID). These pathways converge in the external nucleus of the midbrain (IC) to form a map of auditory space. The map contains space-specific neurones, which respond selectively to sounds coming from specific directions in space or specific combinations of ITD and IID. These neurones do not respond to uncorrelated pairs of sounds. They can detect small amounts (0.3) of correlation between signals going to the two ears. The responses of these neurones to partially uncorrelated signals can account for the behavioral results mentioned above.

Masakazu Konishi
Caltech Pasadena
California, USA

Superior olivary complex

Three main nuclei are part of this complex: the superior lateral, medial, and trapezoid body nuclei. All of them exhibit tonotopy, i.e., the tonal distribution established at the cochlea is maintained. A group of scattered nuclei surrounding the olive, called periolivary, is the origin of the olivo-cochlear efferent fibres. Three different pathways provide the peripheral input to the olive and the lateral lemniscus nuclei: the ventral, medial, and dorsal strias arising from the CN.

The local-field evoked potentials exhibit a clear difference when stimulated ipsi- or contra-lateral to the recording electrode position (Fig. 1.3(A)). Most of the lateral superior olive neurones are excited by ipsilateral sound stimulation (E) while inhibited by contralateral sounds (I), called EI units (Fig. 1.6). Space specific neurones in the owl's auditory map of the space involve computations that determine the interaural time (ITD) and interaural level (ILD) that define the auditory space. The neurones studied behaved like analog AND gates of ITD and ILD, suggesting the two inputs are multiplied instead of being added (Peña and Konishi, 2001). The same neurotransmitters found in the CN are also present in the superior olive complex. The binaural hearing process begins precisely here, at this complex. The main cues the brain uses to localize acoustic information from the environment are interaural intensity and interaural timing differences.

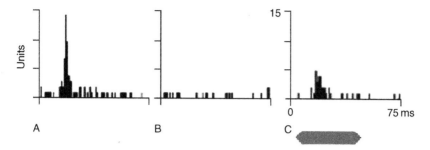

FIGURE 1.6   Awake guinea pig. Post-Stimulus Time Histograms of a lateral superior olive neurone: (A) ipsilateral sound stimulation showing a high probability of discharge at the beginning of the stimulus; (B) contralateral ear stimulation produces no response; (C) the binaural sound stimulation provokes a lower discharge in comparison with the ipsilateral one Tone-burst, 50 ms.

The nuclei of the lateral lemniscus

The division into ventral and dorsal lateral lemniscus is consistent with the presence of two distinct functional systems, a monaural *ventral* (VLL) and a binaural *dorsal* (DNLL) lateral lemniscus system.

The VLL consists of groups of neurons embedded within the lateral lemniscus, located between the superior olivary complex and DLL. It receives inputs mainly from the contralateral ear, as opposed to the DLL which receives inputs from both ears mostly through its connections with the superior olive.

VLL neurons exhibit a variety of shapes and sizes in Nissl stained sections. Most of them project to the ipsilateral IC in laminar fashion. Although previous studies in cat claimed a lack of orientation of the VLL cells, it has been recently demonstrated that the VLL is organized in a laminar fashion suggesting that it is also tonotopically organized. Based on the morphology of the dendritic arbors, two main types of VLL neurons have been described: bushy and stellate cells. The bushy cells show an onset firing pattern and non-linear current–voltage relationship. In contrast, the stellate cells show a linear current–voltage relationship, but exhibit different firing patterns which may be related to differences in the shape of the soma, and the dendritic branching pattern and orientation. Similar firing patterns have also been found in *in vivo* studies.

The *afferent projections* to the VLL arise mainly from the contralateral ventral CN and ipsilateral trapezoid body. The majority of cells in the ventral part of the complex are glycine and/or GABA immunoreactive, although the incidence of co-localization of the two transmitters seems considerably higher than previously estimated. The VLL neurons are suitable for encoding temporal events.

The DLL is a distinctive group of neurons embedded within the dorsal part of the lateral lemniscus. In contrast to the VLL, the DLL receives input from both ears, and it projects to both ICs as well as to its counterpart on the opposite side through the commissure of Probst's. The DLL plays an important role in binaural processing, i.e., sound localization.

Several neuronal types have been described depending on the species and the criteria used for cell classification. Regardless of the morphological type, all cells have similar membrane properties with a sustained series of regular action potentials produced by the injection of positive current.

Generally speaking, the DLL receives a copy of the afferents projections that also innervate the IC. Thus, DLL receives contralateral inputs from the ventral CN and DLL, ipsilateral input from the medial superior olive, superior paraolivary nucleus, and VLL, and bilateral inputs from the lateral superior olive. The DLL, in turn projects to the IC in a laminar and bilateral fashion, with a predominant projection to the contralateral IC. Most DLL cells are GABAergic and therefore have an inhibitory influence on the IC.

Recent studies, based on the blockade of excitatory responses in DLL with kynurenic acid or lesion due to DLL with kainic acid, demonstrated that this causes changes in the responses of neurons in IC to interaural time and intensity differences and reduced the binaural suppression in the IC.

Manuel S. Malmierca, Miguel A. Merchán
Universidad de Salamanca
Salamanca, España

Inferior Colliculus

The key anatomical location of the IC is suitable for interactions among different possible connections, i.e., the lower auditory nuclei and cortex, reticular formation, periaqueductal gray, superior colliculus, etc., as well as among diverse physiological states such as the sleep–wakefulness cycle. An IC role may be to serve as a synaptic crossroad of ascending and descending information, a true auditory processing-integrative function that also depends on the general state of the CNS.

Three sub-nuclei constitute its internal structure: the central, external, and dorsal nuclei (Huffman and Henson, 1990; Oliver and Shneiderman, 1991). The central one is a main station for all ascending information, organized in layers associated with a tonotopic distribution, with low frequencies at dorsal *loci* while higher frequencies are located in ventral regions (Semple and Aitkin, 1979; Oliver and Morest, 1984).

Animal recordings of IC local-field potentials and far-field potentials recordings in humans exhibit some characteristics corresponding to the local multi-unitary activity and the negative wave converging on the IC region, respectively. Neurone unitary firing recordings may be classified using the temporal distribution over the PSTH (Fig. 1.7) (Morales-Cobas et al., 1995).

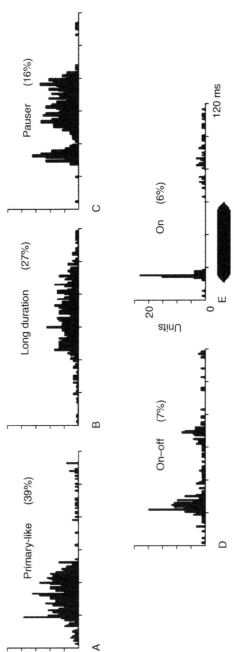

FIGURE 1.7 PSTH from IC units in awake guinea pigs. The recordings, carried out with glass micropipettes and no drugs, demonstrate five discharge patterns: (A) primary-like, (B) long lasting, (C) with notch, (D) on–off, and (E) on.

Intracellular *in vivo* recordings of physiologically identified IC central nucleus auditory neurones and their sub-threshold membrane potential activities were recorded (Pedemonte et al., 1997). The spontaneous action potentials were divided into two groups according to their duration and mean firing rate. Current injection revealed adaptation and membrane potential changes outlasting the electrical stimuli by 20–30 ms. Sequences of synaptic potentials, longer than the sound, were observed lasting up to 90 ms with binaural stimulation. The data from Pedemonte et al. are consistent with the existence of a multi-synaptic pathway by which signals arrive at the IC central nucleus, including corticofugal pathways. This may contribute to the long duration post-synaptic potentials.

The IC excitatory neurotransmitters are amino acids and the inhibitory ones are GABA and to a lesser extent glycine (Faingold et al., 1991). In addition, the IC receives noradrenergic terminals from the *locus coeruleus* cells.

In anesthetized guinea pigs preparation there is little spontaneous activity whereas during W, all IC neurons analyzed showed a high rate of spontaneous firing and a great number of membrane potential oscillations (Torterolo et al., 1995).

Most IC cells are binaurally sensitive (Caird, 1991). The binaural cues are also the ITDs and the IIDs. The characteristic firing of such units led to the creation of a space map in the IC equivalent of the barn owl (Knudsen and Konishi, 1978). This means that particular neurones may respond to stimuli coming from particular regions of the surrounding space, thus creating an auditory map of the world.

Medial geniculate body

This thalamic nucleus is divided into three regions: ventral, dorsal, and medial. The afferents to the medial geniculate body (MG) come from the IC and the reticular formation while its efferent fibers reach the auditory cortex. The ventral portion exhibits a laminar structure and maintains a net tonotopic organization. Each cell of the lamina receives input from the IC which also receives descending fibers from cortical origin (Winer, 1991).

The unitary activity of the MG is similar to that of the IC with inhibitory lateral bands. A great proportion of the cells in the MG are binaural, responding to the ITDs while others are predominantly sensitive to the IIDs. Axons from the thalamic reticular nucleus cells, all of which are GABAergic, also terminate in the ventral nucleus (Montero, 1983) while some sparse cholinergic immunoreactive axons occur in the same nucleus with unknown source (Levey et al., 1987). A specific auditory reticular nucleus (Fig. 1.12) is located in the reticular thalamus contributing to the information transfer to the cortex (Guillery et al., 1998).

Auditory cortex

The human auditory cortex (AI) lies in the superior temporal plane; Heschl's gyrus exhibits a Nissl staining revealing a seven-layer organization originally shown by Ramón y Cajal (1952) (Fig. 1.8).

Tonotopy is also exhibited by the cortex. A general view of cortical physiology is a tonotopic core surrounded by a belt of tissue with a less clear tonotopic representation and flanked by areas that show no evidence of tonotopy at all (Fig. 1.9). The direction of the tonotopic gradient in AI is defined caudomedially for high frequencies while low frequencies are found rostrolaterally (Semple and Scott, 2003). Neurones located in the medial geniculate body provide input to all auditory cortices. The thalamo-cortical projections are the basis for a particular sequential processing through the core, belt, and parabelt auditory areas. Then, auditory signals may be distributed to parietal, temporal, and frontal multimodal areas.

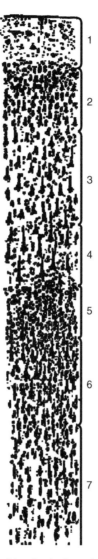

FIGURE 1.8 Nissl staining histological study by Ramón y Cajal described seven laminae in the human auditory cortex. Plexiform layer (1). Two granular laminae (2, 5) and the 3, 4 laminae showing a different pyramidal neurone sizes. Other authors join 6, 7 laminae into only a single lamina 6 (modified from Cajal (1952)).

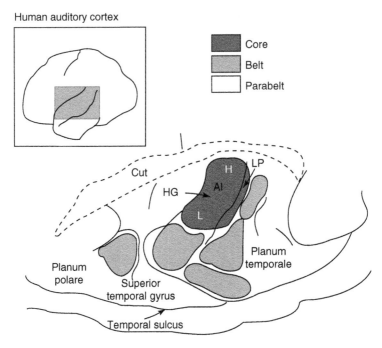

FIGURE 1.9   In humans the auditory cortices lie on the superior temporal plane. HG: Heschl gyrus; LP: lateroposterior area; AI: auditory primary area; H: high frequency; and L: low frequency (modified from Semple and Scott (2003)).

The "what/where" distinction has gained support from animal and human studies (Semple and Scott, 2003). The "where," spatial processing, implicates the dorsal pathways linking the caudomedial belt to frontal and parietal targets. The "what" – objects, phonemes – originates in the anterior core and belt areas reaching targets in the temporal lobe. Human imaging studies (Zatorre et al., 2002) are more or less consistent with the "what/where" hypothesis while a significant correlation between temporal and parietal cortical responses has revealed the possibility of an important cross-talk between the systems (Alain et al., 2001).

The cat cortical local-field potential recording exhibits a long latency response whose amplitude is dependent on the state of the experimental animal, e.g., anesthetized or awake.

Unitary activity

The neuronal activity recorded in the primary cortical area (AI) in awake guinea pigs and monkeys, in response to pure tone stimulation (Fig. 1.10), represents an approach to the central processing of sound by each auditory neurone integrated in a neural network(s). The different PSTH shown, i.e., the probability of discharge over time, are not fixed for each neurone but change their characteristics depending on several factors, e.g., the stimulus intensity, the stimulus spectrum (e.g., noise or tone-bursts), CNS awake or asleep (Peña et al., 1999; Recanzone, 2000).

In our experimental results with guinea pigs during W and sleep never encounter a bursting neurone. A burst is a group of action potentials with 4 ms interspike interval preceded by a silent period of about 100 ms. It was reported that auditory thalamus and also cortical neurones bursts occur in synchronized electroencephalogram states, slow wave sleep (SWS) in a low proportion and under anesthesia. During W someone may appear although in a very low frequency being the greatest presence under anesthesia (Massaux and Edeline, 2003; Edeline, 2005).

Tuning curves

Similar to the auditory nerve curve shown in Figure 1.2 (right), several different tuning curves have been described from the auditory cortex. Single-peaked tuning curves are narrow, excitatory, and V shaped, in response to pure tone stimuli. Others are scarcely tuned, exhibiting wide curves or two peaked and

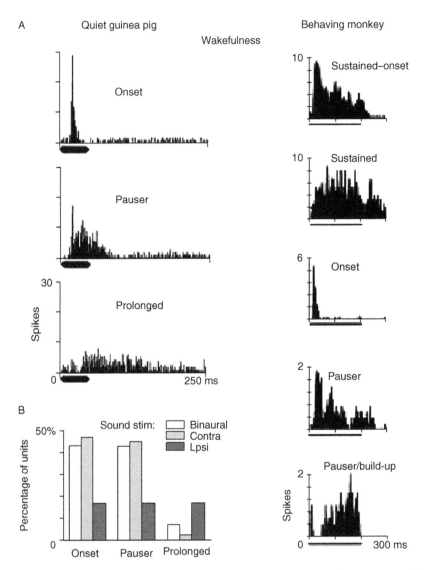

FIGURE 1.10 Left: Neuronal response patterns in the AI auditory cortex of the awake guinea pig. (A) PSTH of the three main response types obtained for contralateral best frequency 50-ms tone-bursts stimuli during W. (B) Response type distribution for monaural ipsi–contralateral and binaural acoustic stimulations (modified from Peña et al. (1999)). Right: Responses for neurons of a behaving macaque monkey classified into five categories. Each post-stimulus time histogram shows eight trials near the characteristic frequency of different AI neurons (modified from Recanzone (2000)). The differences may be due to species differences, animal condition such as quiet W versus behaving, stimulation time, the method of stimulus delivery, closed versus open field, etc.

wide curves, without exhibiting a net characteristic frequency (Schreiner et al., 2000).

Magnetoencephalography

This method is characterized by its very high temporal resolution, which is on the order of milliseconds, as compared with neuroimaging techniques (temporal resolution on the order of minutes). Magnetoencephalography (MEG) has advantages over electroencephalography (EEG) for detecting cortical dipole localization because there is less effect from current conductivity caused by cerebrospinal fluid, skin, etc., while the spatial resolution for MEG evoked activity is on the order of a few millimeters. Thus, the dipole generated may be detected with high precision. During W a multi-wave response to pure tone stimuli was observed in the human primary auditory cortex with waves at 50, 100, 150, and 200 ms (Kakigi et al., 2003; see Chapter 5).

The hippocampal theta rhythm–auditory unit relationship

Phase locking between auditory neurones firing and hippocampal theta rhythm (Hipp θ) during W at different pathway levels, e.g., IC and primary cortex has been demonstrated (Pedemonte et al., 1996a; 2001; Velluti et al., 2000; Velluti and Pedemonte, 2002; Pedemonte and Velluti, 2005b). We have postulated that the Hipp θ could play a role as an internal clock adding a temporal dimension to the auditory processing at auditory pathway and cortical *loci* (see Chapter 5).

Imaging

Functional imaging of cerebral cortical activity relies on coupling of blood flow to neuronal firing and metabolism with functional magnetic resonance imaging (fMRI). The relationship between the hemodynamic fMRI-based signals and neuronal firing is a matter of discussion. Mukamel et al. (2005) compared the human auditory cortical single-unit activity (two recorded patients) with the fMRI of 11 healthy people analyzed in response to identical auditory stimuli. A highly significant correlation between single-unit activity and the fMRI results was obtained, meaning that the fMRI may provide reliable reflection of the firing of human auditory neuronal firing.

The extent of fMRI activation in the superior temporal gyrus increases with stimulus intensity, revealing that intensity lower than 60 dB SPL would make activation impossible to detect. The stimulus frequency also has an important impact: fMRI activation is greater with a stimulus at 1 kHz and stepped tones than with single pure tones. Several stimuli have been used, such as tones, words, etc. Music is the stimulus evoking the largest activation of the primary and secondary as well as associative cortical areas (Fig. 1.11).

Neurotransmitters

The principal cortical neurotransmitter is GABA known from the cortical supra-granular zone in AI. The use of iontophoretic GABA injections demonstrated an inhibition on neurones responding to sound. Noradrenaline has also been described as an inhibitor, although with a slower time course than GABA. Cholinergic cortical neurones inhibit about half of the neurones while the other half is facilitated by acetylcholine (Ach). The cholinergic actions on AI have a long latency and a long duration of several minutes. Some of these actions are blocked by atropine.

The analysis of neurotransmitter roles in the auditory pathway, particularly in the IC in sleep and W can also contribute to the understanding of neural information processing. The

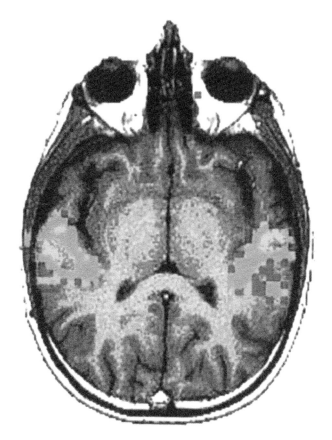

FIGURE 1.11 Auditory functional MR imaging with music. The intensities of brain regions activation varied from high (yellow) to low (blue). The primary auditory cortex is bilaterally activated while the activity of secondary auditory cortex is greater on the left (modified from Bernal and Altman (2001)). See Plate 1.11 for the colour version of this figure.

results reported address the afferent and efferent actions in which excitatory neurotransmission is involved. With respect to the N-methyl-D-aspartate action on IC cells, it can be concluded that there are no differences, at the cellular level, between sleep and waking (Goldstein-Daruech et al., 2002). Thus, other types of excitatory neurotransmission, as well as synaptic inhibition, may underlie the physiologically complex changes that occur in the sleeping brain. The excitatory efferent system was postulated as being mainly glutamatergic (Feliciano and Potashner, 1995). Thus, the effect of auditory (AI) electrical stimulation was mimicked, a few minutes later, by iontophoretic kynurenic acid ejection onto the same IC neuron, supporting the notion of an efferent pathway acting through excitatory amino acid receptors on IC cells and/or on inhibitory neurones (Goldstein-Daruech et al., 2002; Velluti and Pedemonte, 2002). Thus, the descending auditory system can participate in the processing of the incoming information. Excitatory amino acid transmission would be present in sleep and W acting by similar mechanisms together with several other neurotransmitters on the same neurone.

The reticular formation

Also receiving sensory information through collateral fibers of the sensory pathways in general and the auditory pathway in particular (Huttenlocher, 1960), the reticular formation is an important structure because of the general awakening role that it may play (Moruzzi and Magoun, 1949). From the functional point of view, this region is part of a diffuse spreading of information as well as orderly information flow toward the superior processing centers. Moreover, there are specific sensory nuclei located in the thalamic reticular formation, including a well-defined reticular auditory nucleus; this particular auditory reticular region is connected to other thalamic regions and to the auditory cortex (Fig. 1.12).

The cerebellum

The cerebellum constitutes another auditory processing center whose function is still under research. The auditory information reaches the cerebellum as evidenced by evoked potentials

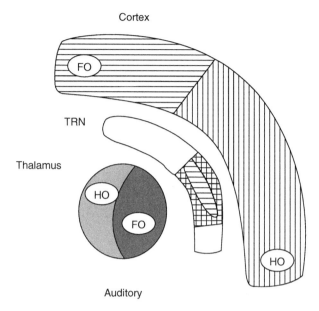

FIGURE 1.12 Thalamic reticular nucleus (TRN). Three reticular layers can be recognized in the auditory region of the TRN. The central TRN layer receives from FO, called first-order nuclei, and their main input from ascending specific fibers and the cortical horizontal stripes. Inner and outer layers receive input from higher-order thalamic and cortical regions, vertical stripes. HO, called higher-order nuclei, receive their main driving input from cortical layer V (modified from Guillery et al. (1998)).

recorded, e.g., at the rat vermis (Lorenzo et al., 1978) and the evoked resistance shifts in cats (Galambos and Velluti, 1968). In coordination with the visual and somesthetic neuronal activity of the superior colliculus, this cerebellar auditory information may underlie functions such as the control of head movements during the search of a sound source in space and perhaps also in learning. Cerebellar vermis electrical stimulation introduces changes in the cochlear nerve and microphonic activity, thus contributing to efferent system functions (Velluti and Crispino, 1979).

Then, it has been established that the evoked potentials, the magnetoencephalographic evoked activity, and the unitary neuronal firing – recorded at different loci of the auditory pathway – exhibit changes related to different physiological brain states. For instance, the sleep–wakefulness cycle, the habituation process, learning, etc. Finally it can be assumed that all the changes mentioned are due to the interactions with other auditory system components, e.g., the complex and powerful efferent system.

The efferent descending system

There are several mechanisms by which the CNS controls and/or modulates the incoming auditory signals. I will mention the following examples: (a) ears' movement (mainly in lower mammals), (b) movements of the middle-ear muscles, (c) regulation of the mechanical properties of the outer and inner hair cell of the organ of Corti, (d) actions over the auditory nerve primary afferents fibers, and (e) actions on each nucleus of the auditory pathway.

How is the efferent system organized?

There are two opinions about this subject. One supports the notion of three interconnected feedback circuits: (1) cortex–thalamus–mesencephalon; (2) mesencephalon–superior olivary complex–cochlear nucleus; (3) superior olivary complex–cochlear nucleus–cochlea. The second position supports the idea of a continuous descending chain with actions and feedback loops at all levels (Spangler and Warr, 1991). An important experimental datum supporting the second position is the possibility of producing similar effects to those which activate the olivo-cochlear bundle through stimuli in the higher regions of the auditory system (Desmedt, 1975). Due to the complexity of the system, it becomes difficult to support a single position because both possibilities can coexist (Fig. 1.13).

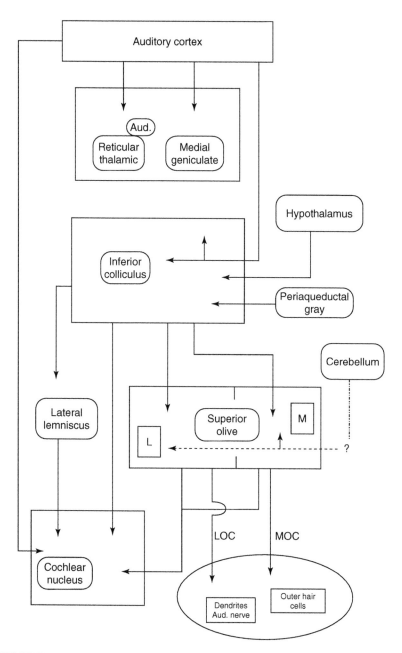

FIGURE 1.13 General diagram of the efferent (descending) pathways.

Moreover, signals originating in the cerebellar vermis have been reported to modify the auditory cochlear responses, thus contributing to the central control of this sensory input at the receptor level (Velluti and Crispino, 1979). The periaqueductal gray has also been shown to have an influence on (as well as an anatomical connection) to the CN unitary activity (Pedemonte et al., 1990; Radmilovich et al., 1991).

Auditory cortex efferents

Two outgoing cortical systems are described: one connected to the medial geniculate nucleus and the other, with a wider distribution, is projected toward the IC, the non-auditory thalamic nuclei, the *striatum*, and the lateral pontine region (Spangler and Warr, 1991).

Auditory corticofugal fibers constitute a descending parallel pathway reaching the MG monosynaptically (Winer et al., 2001), the IC (Winer et al., 1998), the CN (Schofield and Coomes, 2005), and the olivo-cochlear neurons (Mulders and Robertson, 2000). The axons descending from the cortex to the IC terminate in the different regions of this nucleus including the central one (Faye-Lund, 1985; Saldaña and Merchán, 1992; Saldaña et al., 1996), showing the descending fibers in the IC innervate wide zones which go beyond the specifically auditory central nucleus and the contralateral IC.

Scarce data have been published about the effects of the auditory cortex on the neuronal activity of the medial geniculate nucleus (Amato et al., 1969). Actions of the electrically stimulated primary auditory cortex on the medial geniculate have been reported by Sauerland et al. (1972). The authors showed the existence of presynaptic excitability increase in this nucleus, a phenomenon which is capable of modulating the auditory input to the cortex.

Little is known about the physiological aspects of the dorsal efferent system. Both excitatory and inhibitory potentials in neurones of the inferior colliculus in response to ipsilateral electric stimulation of the auditory cortex have been reported (Mitani et al., 1983). The same was demonstrated for the central and dorsal nucleus of the inferior colliculus with extracellular recordings (Syka and Popelar, 1984; Sun et al., 1989). The existence of bilateral cortical actions on the activity of the same IC auditory neurone has been described (Torterolo et al., 1998). These results suggest the possibility of more complex processes which would include the integration of the ascending information plus the efferent actions of both cortices on the same collicular neurone. Moreover, firing shifts may be obtained by stimulating the AI as well as other cortical regions such as the orbital gyrus (Sauerland et al., 1972; Sun et al., 1989; Torterolo et al., 1998).

This excitatory descending system was postulated as being mainly glutamatergic (Feliciano and Potashner, 1995), whereas cortical electrical stimulation affects the ICc neurons decreasing the firing of 86% of the recorded neurones and increasing the firing in the remaining 14% (Fig. 1.14). Approximately, the same proportions were obtained when, with a double micropipette, an NMDA blocker was electrophoretically injected on the same ICc unit (Goldstein-Daruech et al., 2002).

Inferior colliculus efferents

This nucleus has multiple connections to the cortex, which provides high-capacity parallel channel for processing auditory information. Evidence for extensive and reciprocal hypothalamic and IC connections have been recently reported (Winer, 2006). From the IC, descending fibers project to the superior olive and the CN (Huffman and Henson, 1990; Saldaña et al., 1996). The projection to the superior olive ends in the neurones that originate in the olivo-cochlear system, so, serving as

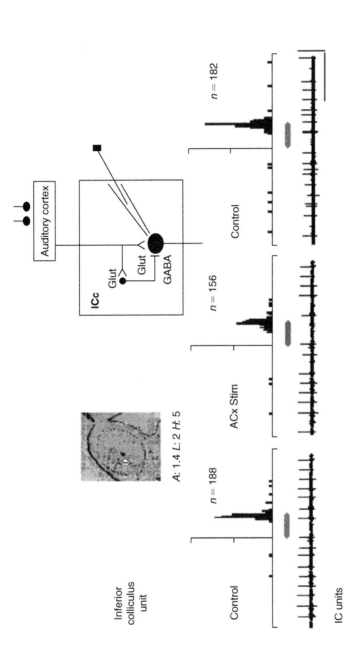

FIGURE 1.14 Effect of auditory cortex (AI) on an auditory IC neurone. PSTH show that cortical electrical stimulation produce decrement in the response to sound and the possible GABAergic influence. ICc: inferior colliculus central nucleus; ACx Stim: auditory cortex stimulation; Glut: glutamate; GABA: gamma amino butyric acid; *n*: number of spikes; sound stimuli: contralateral tone-burst, 50 ms duration at the unit characteristic frequency; 70 dB SPL (modified from Goldstein-Daruech et al. (2002)).

a connection from higher centers to the most peripheral regions of the efferent system (Faye-Lund, 1986).

The olivo-cochlear system

The olivo-cochlear efferent neurons originate in the brainstem and terminate in the organ of Corti, thereby allowing the CNS to control cochlear and auditory nerve function. This bundle of fibers, first observed by Held (1893) and Lorente de Nó (1933) in connection with the CN, has been extensively described by Rasmussen (1946; 1960). Primarily, the efferent fibers split into one crossed bundle and one direct bundle although nowadays it is agreed that there exists (a) a medial system and (b) a lateral one (Warr, 1975; Guinan, 1986). The medial system, whose neurones are located in the superior medial olive on either side, sends its axons to the outer hair cells. The lateral system, with its neurones located in the periphery of the lateral superior olive – ipsi- and contralateral – makes synapse with the afferent auditory fibers beneath the inner hair cells. Another important anatomical difference between the medial efferent fibers and the lateral ones is that the former are myelinated, while the latter are unmyelinated (Warr, 1992).

ACh is the main neurotransmitter released by the medial olivo-cochlear (MOC). The lateral olivo-cochlear (LOC) is also cholinergic although it is now suggested that LOC fibres co-localize many neurotransmitters and neuromodulators. Recently it was proposed that the efferent innervation of outer hair cells is mediated by the heteromeric $\alpha 9$ and $\alpha 10$ subunits of the nicotinic ACh receptor (Elgoyhen et al., 2001).

Physiological approaches

Galambos (1956) experimentally demonstrated the capacity of the olivary-cochlear bundle to reduce the amplitude or to block the cAP of the auditory nerve by electrically stimulating the direct and crossed bundles of efferent fibers at the floor of the IVth ventricle. The evoked decrease of the cAP is proportional to the intensity of the electric stimulation and to the stimulating sound. Subsequently, Fex (1962; 1967) using single-unit recordings, reported that such effects corresponded to an inhibitory process and added the observation that the cochlear microphonic (CM) increased in amplitude in response to electric stimulation of the floor of the IVth ventricle.

On the other hand, after systemic injection of benzodiazepines, we reported an inverse change (Velluti and Pedemonte, 1986): while the cAP increased the CM diminished its amplitude. Then, we postulated a central action of benzodiazepine on GABA-ergic neurones, probably onto efferent periolivary neurones, that would determine the actions on both the receptor (CM) and the auditory nerve (cAP). The experimental control – the peripheral injection of the same benzodiazepine – did not have any effect when applied directly to the cochlea.

The conclusion is that all possibilities are present in the system, i.e., the cAP and the CM amplitude may increase and/or decrease, depending on the way the system may be activated. Different experimental approaches, exhibited in Table 1.2, produce all the possible changes in both potentials amplitudes.

Relationships with attention processes

Experimental approaches demonstrated that cochlear recorded potential amplitudes may also shift in a parallel fashion. Under different physiological conditions such as habituation and dishabituation to sound (Buño et al., 1966) and during sleep (Velluti et al., 1989), the cAP and the CM amplitudes change in the same way, the two simultaneously recorded potentials increasing or decreasing. Accordingly, Oatman (1971; 1976) provides information correlating the activity of the efferent system

TABLE 1.2 Auditory nerve action potential and cochlear microphonic amplitudes shifts, correlated with diverse experimental conditions.

Authors	Experimental conditions	cAP shifts	CM shifts
Galambos (1956)	Anesthetized cats. Electrical stimulation on the floor of the IVth ventricle.	⌇ ▲ ⌇	
Fex (1962a)	Anesthetized cats. Electrical stimulation on the floor of the IVth ventricle.	W ▲ W ⌇ ⌇	W ▲ W
Buño et al. (1966)	Awake behaving guinea pigs. Habituation to constant stimuli. Without ossicles and closed sound stimulating system.	W ▲ W ⌇ ⌇	▁▁ ▲ ▁▁
Oatman (1971)	Awake cats without middle-ear ossicles. cAP and changing attention with visual stimulation.	W ▲ W ⌇ ⌇	
Velluti and Pedemonte (1986)	Awake guinea pigs with systemic injected benzodiazepine. Without ossicles and closed sound stimulating system.	W ▲ W ⌇	W ▲ W

(Continued)

TABLE 1.2 (*Continued*)

Authors	Experimental conditions	cAP shifts			CM shifts		
Velluti et al. (1989)	Awake and sleeping guinea pigs. Without ossicles and closed sound stimulating system.	W	▲	SWS	W	▲	SWS
Pedemonte et al. (2004); Pavez et al. (2006)	Awake and sleeping guinea pigs. Without ossicles and closed sound stimulating system.	W	▲	SWS	W	▲	SWS
Délano et al. (2007)	Awake chinchillas. Cochlear recorded potentials and attention shifts.	W	▲	W	W	▲	W

cAP: auditory nerve compound action potential; CM: cochlear microphonic; W: wakefulness; SWS: slow wave sleep.

to attention processes: when a laboratory animal is performing a visual task, the amplitude of the auditory nerve potential shows an amplitude decrease. Moreover, it has recently been reported that visuo-spatial attention modulates cochlear amplitudes decreasing the cAP and concomitantly increasing the CM amplitudes (Délano et al., 2007) (Table 1.2).

Modifications of the mechanical status of the cochlea

Electromotility, the ability to produce movements such as contractions and stretchings of the outer hair cells (Brownell et al., 1985), permits the control of the displacements of the basilar membrane, thus influencing the cochlear mechanics and, consequently, the transduction process and the sensitivity of the receptor system, e.g., the inner hair cells. This capacity to change the sensitivity of the system is called "the cochlear amplifier." In accordance with the definition of Robles and Ruggero (2001), the "cochlear amplifier" is a positive feedback process which increases the sensitivity to the response of the basilar membrane in response to low-intensity stimuli. Scarce experimental data are available concerning the effects of the efferent system on the mechanical control of the basilar membrane.

Otoacoustic emissions

Acoustic emissions are detectable sounds in the external auditory canal generated by the cochlea (Kemp, 1978). A series of experiments have shown the influences of the medial efferent system on otoacoustic emissions which provide indirect information about the motility of the basilar membrane. Distortion products are tones produced in the cochlea by distortion as a response to stimulation with two primary tones f1 and f2 with different frequencies. The wave generated at the level of the distorting frequency, due to non-linearities, which are probably located in the basilar membrane, expands in both directions from its point of origin at the point of maximum response to f2. Efferent stimulation may decrease the distortion product amplitudes or sometimes increases them (Mountain, 1980; Siegel and King, 1982). The effects of the medial efferent system on otoacoustic emissions seem to indicate a mechanical change in the cochlea. At low stimulus sound levels, depression of the basilar membrane movement by stimulating the medial efferent system appears as the dominant mechanism for modulating of inner hair cells function. This changes the receptor sensitivity and thus can change the afferent fiber discharge rate. The mechanism that could underlie efferent system modification of the cochlear vibration would be the electromotility of the outer hair cells. However, it has been reported that an additional cellular mechanism, which concerns active movements of the cilia, could contribute to increased system sensitivity and could likely also be part of the origin of the otoacoustic emissions (Martin and Hudspeth, 1999).

General considerations of efferent activities

The discrepancy between the natural environmental noise level and the experimental high-level noise used to evoke the MOC bundle activity, evidence that the MOC system did not evolve to protect the ear from natural sound acoustic trauma (Kirk and Smith, 2003). Only in rare instances are ambient noise level sustained at moderately high intensities, e.g., the highest noise in a natural environment corresponds to frog choruses reaching about 90 dB SPL (Narins and Hurley, 1982). By contrast, all experiments in which a MOC-mediated protective effect was observed using a much higher sound intensity, ~100–150 dB SPL. Besides, the MOC is present in all mammals evolved during 170 million years in the total absence of sustained high intensity and narrow frequency natural noise.

The global functions of the efferent system, and particularly the olivo-cochlear partition, are not well known. Hence, the main hypotheses for its function are:

1. To improve the detection of a signal masked in noise.
   The MOC system very possibly evolved for unmasking significant sound reducing simultaneous low-level noise – a hypothesis with wide experimental support.
2. To modify the mechanical status of the cochlea by providing control of the outer hair cells and cilia movements influencing inner hair cell activity and otoacoustic emissions.
3. To control some natural body-produced noises. High-intensity internal noise, e.g., the chewing, respiratory, heart and blood flow noises, etc., shall be controlled to maintain the system in optimal functional conditions. Other physiological noises, e.g., the circulation through the carotid arteries located close to the cochlea, are blocked and not consciously perceived although they are generating continuous acoustic input reflecting body physiology: the auditory nerve activity associated with blood flow noise described by Lewis and Henry (1992) and the CN single-unit firing synchronized with heart beats–blood flow, reported by Velluti et al. (1994).
4. I now offer a new hypothesis regarding efferent fiber function. The efferent system, which is connected to a great diversity of auditory-related regions, exerts actions over the input and the processing at different levels, which means that efferent activity can establish a synergy between the auditory system and the ever-changing CNS status. Synergy means to put both the brain status and the auditory input reciprocal interactions to work in order to support the widely distributed brain changes occurring on entering sleep. This hypothesis is introduced to explain the efferent

actions of all sensory systems, and the auditory system in particular. The continuous interaction/adaptation between the CNS and the outside and inside world (the body) can be achieved only through feedback systems (receptors, afferent and efferent pathways) which are capable of constantly following both rapid and slow changes of the state of the CNS.

Finally, this new key function presented for the auditory efferent system is supported by several experimental approaches (Velluti, 1997; 2005; Velluti et al., 2000; Velluti and Pedemonte, 2002; Pedemonte and Velluti, 2005a,b), particularly those coming from system physiology of both behaving, awake or asleep animals and humans.

2

The physiological bases
of sleep

All living creatures, being plants or animals, unicellular or complexly organized, oscillate in time configuring rhythms which arise as a result of sensory modulation acting on genetically encoded information. By using the electroencephalography (EEG), it is possible to typify each behavioral state: wakefulness (W), stages I and II, slow wave sleep (SWS) with stages III and IV, and paradoxical sleep (PS) with or without rapid eye movement (REM), whose overnight sequence shows a characteristic architecture that constitutes an ultradian rhythm. The physiology of many different systems is modulated by the wakefulness–sleep cycle: the processing of sensory information, the oneiric activity, the cardiovascular and respiratory functions, the endocrine functions, as well as the body temperature control, the homeostasis, and the energetic metabolism, all happen to change depending on the moment of the cycle. There are a number of neural centers and networks involved in the generation and maintenance of this cycle. Much has been speculated about the possible actions of sleep, but little is known about this state essential for life which takes a third of it.

Sleep is a central nervous system state

Along evolution, more and more complex systems have arisen to control the narrow margins of normality. However, a normal range is not constant but presents oscillations and rhythms.

Although it is possible to recognize rhythms of activity and rest in practically all living creatures while sleep, with all its particular physiological characteristics, appears in homeotherm vertebrates. Along phylogenetic evolution, this rhythm is found to be ultradian in most of species (Esteban et al., 2005). Even though the whole central nervous system (CNS) participates in this state, certain entity, such as the basal forebrain are mainly related to the SWS organization, while the *dorsolateralis pontine tegmentum* contributes with characteristic expressions, or signs of the PS. The pons being part of a Final Common Region for PS phenomenology (Velluti, 1988). Sleep arises as an ensemble of physiological changes where different systems take part under the regulation of the CNS, particularly the 85% of the whole system, the cortical mantle.

Polysomnography

Studies of EEG rhythms carried out in humans, permitted to classify four stages of sleep (Loomis, 1938), being PS first described as a separate stage by Aserinski and Kleitman (1953). The polysomnogram is the continuous and simultaneous recording of physiological variables during sleep. The minimum recording for stage identification includes: EEG, electromyogram (EMG), and electro-oculogram (EOG). The particular combinations of the three bioelectrical signals – amplitude and frequency of the waves in the EEG, electromyographic, and oculomotor activity – permit the recognition of the different sleep stages (Fig. 2.1). Thus, both W with its variations and sleep with its stages – SWS or orthodox or non-REM (NREM),

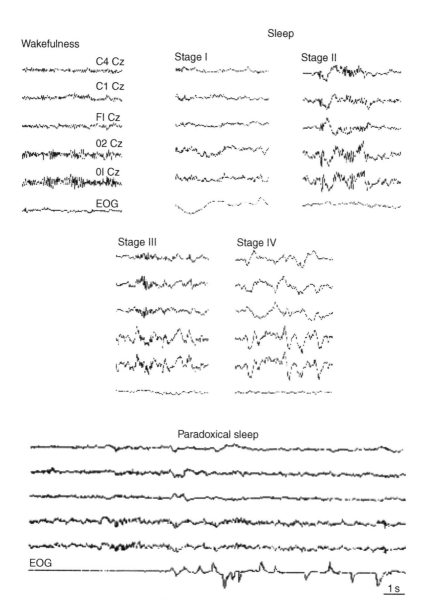

FIGURE 2.1 Brain bioelectrical activity during sleep and waking. The EEG and electro-oculogram (EOG) of a quiet episode of W are shown (note the α rhythm in the EEG). Sleep is divided into stages I–IV, by the characteristics waves, being the last two the classical SWS. The PS represents about 25% of the total amount of sleep and is periodically associated with the presence of REM (EOG).

and PS or active sleep or REM sleep – can be defined using these variables, always associated with behavior. The transition from one sleep stage to another is progressive and the polygraph elements change, with different temporal courses, until achieving completely the next stage characteristics. W is a state in which a characteristic mental activity is developed – the vigil consciousness – accompanied by the execution of voluntary movements. It is not a homogeneous state but it is composed of numerous activity–rest ultradian cycles of about 90-minutes duration.

Since the discovery that electrical stimulation of a mesencephalic region provokes behavioral and electrophysiological awakening in animals, EEG activation, an ascending reticular activating system with a diffuse projection to the forebrain has been demonstrated (Moruzzi and Magoun, 1949; Moruzzi, 1963; 1972). However, this assumption was contradicted by several experimental approaches (Adametz, 1959; Chow and Randall, 1964; Yamamoto et al., 1988; John and Ranshoff, 1996). Two experimental lesions successive in time, separated by days, of the mesencephalic reticular formation do not produce a comatose state; rather, animals are able to feed normally and have sleep–waking cycles. Moreover, they can retain and produce conditioned responses after the second mesencephalic lesion, which indicate that there is not a unique "waking center." Some rostral regions can hold the ability to maintain consciousness even after the mesencephalic lesions (John, 2001; 2006).

Genetic aspects

Recent advances have opened new perspectives in the genetic approach to normal sleep although experimental results are scarce. The findings that certain disorders, such as the fatal familial insomnia, are transmitted genetically and the role of hypocretins in narcolepsy (Chemelli et al., 1999) have opened an avenue for the genetic approach to sleep pathologies.

There are just a few genes whose mutations cause sleep disorders. Furthermore, disorders related to just one gene are rare; neither do we know the influence of environmental factors that may induce those genes expression. The recently identified autosomal recessive gene that controls the frequency of the hippocampus theta rhythm during PS represents an example of genetic regulation of bioelectrical rhythms (Tafti, 2003; Dauvilliers et al., 2005).

Each component of sleep has to be considered as a complex phenotype. Studies in identical and fraternal twins have contributed to determine possible environmental factors. A high similarity in sleep parameters among twins has been reported, thus suggesting an important contribution of genetic factors to the sleep organization (Dauvilliers et al., 2005). However, these studies are scarce and can still be deemed preliminary.

Sleep deprivation

Human volunteers who underwent total sleep deprivation (over periods up to 200 hours) presented signs of great fatigue, attention disorders and irritability, and a significant decrease in discrimination abilities. In some cases they presented hallucinations and balance, sight and speech disorders. The W EEG in sleep-deprived subjects shows a decline in the alpha rhythm: volunteers cannot keep this rest rhythm for more than 10 seconds. Besides, episodes of delta and theta waves become frequent. Selective deprivation of PS cannot be sustained for long, as micro-sleep episodes invade W uncontrollably, whereas total deprivation of 16 hours or longer in humans implies the loss of cognitive abilities, which was well demonstrated experimentally (Durmer and Dinges, 2005). Total sleep deprivation in rats revealed that these animals die within 15–22 days suffering from a general functional depression, presenting neurological and behavioural disorders.

Sleep and their associated physiological changes

During the last decades, it has been demonstrated that all physiological functions vary depending on the moment of sleep–wakefulness cycle, and many of these functions are associated with a particular sleep stage. Thus, the autonomic control expressed through cardiovascular and respiratory manifestations is modified concomitantly with the SWS–PS sequence. The sensory processing, the endocrine function, and others analyzed later in this chapter, are also interrelated with the sleep–wakefulness cycle. Furthermore, during PS the homeostasis is transitorily neglected (Parmeggiani, 1980).

Sensory information processing

Information from the outer world and the body (inner world), both conscious and unconscious, enters through sensory receptors and is continuously evaluated in the CNS. This information connects the individual with the environment and also keeps the brain in close contact with his internal medium, viscera, muscles, joints, and so on (Velluti, 1997). I want to stress now that the information about "inner world" is continuously processed in sleep and waking by the constantly working CNS (Velluti et al., 2000; Velluti and Pedemonte, 2002).

During sleep, psychomotor reactions to environmental stimuli are clearly reduced. We find ourselves relatively isolated from the environment. However, from an electrophysiological point of view, the evoked potentials or the unitary responses of certain neuronal groups are comparatively greater during SWS than during W. This fact, apparently paradoxical, was demonstrated for visual and auditory information. The thalamic and cortical auditory evoked potentials exhibit greater amplitude during SWS when compared to W and PS (Campbell et al., 1992; Coenen, 1995; Velluti, 1997; Bastuji and García-Larrea, 1999; 2005).

The auditory system is a tele-receptor system which remains relatively "open" during sleep. The possibility of establishing auditory contact with the external world must have been important, from a phylogenetic point of view, for the survival of the most vulnerable species, allowing them to wake up and develop an adequate reaction before a predator strikes. At present, we keep using the auditory system to provoke our daily awakening. Much experimental data nowadays support the idea that the entire auditory system, from the cochlear receptor to the cortical neurons, keeps processing information during sleep, although differently from during W. Shifts in the neuronal discharges correlated with sleep stages have been verified along the whole pathway and would possibly be the electrophysiological expression of the changes in sensory processing that take place while we sleep. Thus, we are able to perceive an auditory stimulus, process it and compare it to information stored in our memory and make decisions such as waking up or continue sleeping (Velluti and Pedemonte, 2006).

These results suggest that neurones belong to different neuronal networks which participate in multiple processing besides the specific sensory one, being the general state of the brain, the one in charge of modulating this activity (Pedemonte and Velluti, 2005a,b; Velluti, 2005). Inversely, the complete absence of auditory input produces modifications in sleep and W, increasing both sleep phases to the detriment of W (Pedemonte et al., 1996b).

The sleeping brain imposes conditions for the income and processing of auditory sensory information; furthermore, it is my tenet that what is observed in the auditory system can be valid for all the sensory systems. Thus, sensory activity arriving in the CNS sleep during early development, which in human neonates mostly takes place during sleep, is a relevant factor in the "sculpting" – or maturing – of the brain. The sensory information processed during sleep in an early stage of life (days, months) must participate in the CNS maturation, since sensory information keeps entering while we sleep.

Sleep and dreams

Current physiological studies demonstrate that dreams are regularly present in all phases of sleep, stage II, SWS stages III–IV, and PS, reflecting a more or less complex series of sensory and emotional events (Portas, 2005). However, dreams that take place during the last cycles of PS, just prior to awakening, would be the easiest to recall and thereby the most frequently reported. It is possible to obtain reports of dreams in 85–90% of the awakenings provoked during PS and in 50% of those provoked during stage II and SWS stages III–IV (Foulkes, 1962; Cicogna et al., 1991).

The idea that PS dreams are bizarre, unreal, hallucinating, etc., seems to result from the scarce amount of studies on which this tenet was based. Properly controlled analyses performed in the sleep laboratory show that during PS there are also dreams associated with common experiences of daily life. Recent researches state that the content of those dreams recorded in SWS and PS are equal as long as they have equivalent duration. These concepts lead to postulate the existence of one single dream generator system, which functions throughout the diverse sleep phases. Besides, it is important to remark the coherence and thematic organization of every dream, which reveals that dreams are the product of a brain working in an organized fashion. Some authors have divergent viewpoint to the proposed dichotomy between PS and stage II and SWS dreams, sleep mentation (Hobson et al., 1998). Other researchers consider the differences as quantitative instead of qualitative (Cicogna et al., 1991).

Concerning the thematic content of dreams, 100% of the narrations report visual images and 65% report auditory sensations (McCarley and Hoffman, 1981; Sante de Santis, 1899), whereas the percentages associated with other sensory modalities are significantly lower. Depending on the sensory system involved, information from the environment can "enter" a dream to become part of the report. The importance of the auditory system, which

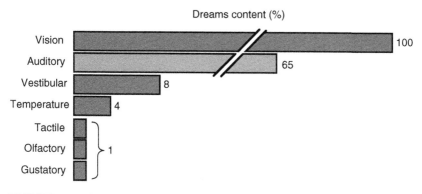

FIGURE 2.2   Sensory modalities present in the dreams contents. The dream visual images (100%) together with auditory "images" (65%) are the most salient components of dreams (modified from McCarley and Hoffman (1981)).

remains constantly "open," lies on the possibility to continuously control the environmental sounds (Fig. 2.2). Although less frequently than sensations, emotions are also expressed in dreams, anxiety being the most often reported. Dreams can repeatedly produce changes in heart and respiratory rates as a result of the autonomic system activation.

The motor activity is also present during dreams although significantly reduced by the motor neurones inhibition present during sleep (Pompeiano, 1967; Lai and Siegel, 1991). The spinal motor neurones inhibition in PS opposes to the increased bursting firing of pyramidal tract fibers (Evarts, 1964).

Although the specific sources of dreams remain an enigma, we can generally state that they are constituted by: (a) what the individuals have in their memory; (b) the sensory information from the outer and inner world – unconscious – intrusions while the dream is taking place; and (c) possible genetically transmitted information.

The fact that depressed patients have dreams with depressive characteristics and schizophrenic patients develop disorganized dreams point to a continuation between the psychical activity of W and the oneiric activity. But there is at least one

objective difference: dreams focus on only one oneiric experience, while during W it is possible to sustain multiple elements
consciously and simultaneously.

Dreams were also interpreted as a rehearsal for brain activity
involved in relevant behavior (Cartwright, 1974). However, it was
also postulated that dreaming may be an accidental by-product
of brain activity; besides, Jouvet (1999) proposed that during
dreaming, the brain may be genetically "reprogrammed."

Cardiovascular functions

During sleep, the mean arterial pressure decreases as a result
of a drop in the diastolic and systolic blood pressures. The lowest value is recorded during stages III–IV of SWS. In humans,
the PS blood pressure becomes variable and exhibits transient
increases of up to 40 mmHg which overlap with a tonic hypotension. These blood pressure increments coincide with the
phasic events (REM, muscle twitches) of PS. The pulmonary
artery blood pressure remains stable during all sleep phases (see
Silvani and Lenzi, 2005).

The heart rate decreases during SWS, predominantly in
stages III–IV. During PS it becomes variable mainly during
the phasic activity. The cardiac output becomes moderately
reduced during both SWS and PS, constituting another element
that contributes to the mentioned blood pressure drop. Sleep
hypotension also depends on vasodilation. During PS, there are
periods of vasoconstriction in the skeletal muscles, which may
be the cause of the phasic increments in blood pressure.

Cerebral blood flow

Studies in human subjects using diverse techniques have
revealed regional raises or drops in blood flow during SWS.
Studies in humans and other species agree in the existence of

a significant rise in cerebral blood flow (CBF) during PS, with phasic increases overlapping the tonic increase (Greenberg, 1980). The mechanisms responsible for these changes have not been clearly characterized, although existing data suggest that they might be induced by local metabolic variations (Reivich, 1974; Braun et al., 1997).

The recording of cerebral oxygen in cats exhibits a particular distribution of O_2 availability during PS, which has been denominated "pO_2 PS system" (Velluti, 1985). This system includes the reticular formation, the *reticularis pontis oralis*, the basal forebrain, the hypothalamus, the amygdala, and the cerebellum. Oxygen availability in the cortex does not show changes during PS, presumably because its blood supply has a better distribution than subcortical structures or because of the cortical laminar arrangement could prevent the recording of the pO_2 oscillating pattern (Fig. 2.3).

This oscillating pattern, which were attributed to a decrease in the local homeostatic control, have also been interpreted as an increase in glucose degradation induced by anaerobic metabolism during neural activity increments (Velluti et al., 1965; García-Austt et al., 1968; Velluti and Monti, 1976; Velluti et al., 1977; Velluti, 1985; 1988; Franzini, 1992).

Neurogenic vasomotricity has also been involved in the CBF regulation, although this aspect has not been sufficiently studied so far. The most widely accepted nowadays is the coupling of blood flow and neuronal activity, by metabolic factors. Experimental data of the last decade, which show a global reduction in the CBF during the night with flow values post-sleep significantly lower than flow values presleep, would support the classical hypothesis of the "restorative" function of sleep (Zóccoli et al., 2005). Once sleep requirements have been fulfilled, resuming the operational brain level would take place at a lower metabolic cost during W subsequent to sleep. Throughout W, a "debt of sleep" would gradually be created, manifested as a growing propensity to sleep, which would be "repaid" during sleep. Since

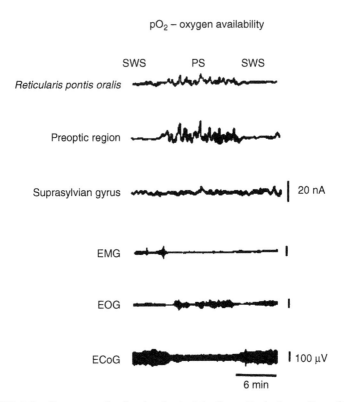

FIGURE 2.3   Oxygen cathodes implanted in the *reticularis pontis oralis*, the preoptic region, and the suprasylvian gyrus recorded local oxygen availability during SWS and PS, in cats. The upper two traces show high-amplitude oscillations, which were not observed in the cortex. ECoG, electrocorticogram.

SWS stages are longer at the beginning of the night and decline exponentially in the course of the hours with a temporal course similar to that of the CBF, some authors speculate about a functional association between both, i.e., the CBF diminishes over the night and slow waves gradually declining.

Imaging

Studies performed in humans using positron emission tomography (PET) have opened a new avenue for the research of sleep processes. Maquet (2000) has showed, through the use of PET with *2-deoxy-D-glucose*, a 12% decrease of the cerebral glucose metabolism during SWS compared to W, while PS produced a general increase of 16% (Fig. 2.4).

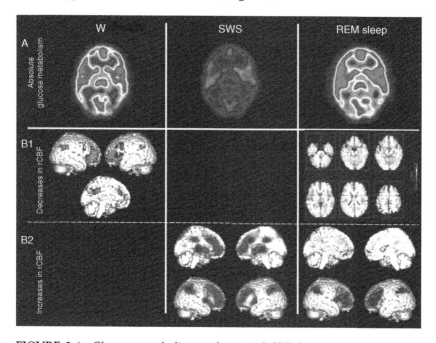

FIGURE 2.4   Glucose metabolism and regional CBF during W, SWS, and PS (REM). (A) Cerebral glucose metabolism quantified in the same individual at 1-week intervals, using fluorodeoxyglucose and PET. There is a significant decrease in the average glucose metabolism during SWS compared to W. During PS (REM) the glucose metabolism is as high as during W (Maquet et al., 1990). (B1) Distribution of the highest brain activity, assessed by CBF measurement using PET during W and PS (REM sleep). The most active regions during W are located in the associative cortices in the prefrontal and parietal lobes (Maquet, 2000). During PS (REM), the most active areas are located in the pontine tegmentum, the thalamus, the amygdaloid complexes, and the anterior cingulate cortex (Maquet et al., 1997). (B2) Distribution of the lowest regional brain activity during SWS and PS (REM sleep) using the same method as in B1. In both sleep stages, the least active regions during W are located in the associative cortices in the prefrontal and parietal lobes. During SWS, the brainstem and thalamus are particularly deactivated (modified from Maquet et al. (2005)). See Plate 2.4 for the colour version of this figure.

Experimental evidence demonstrates that cortical activity, in addition to participate in the global changes that characterize the electrophysiological configuration of sleep, is also involved in specific processes; it is reactivated during the SWS and PS episodes subsequent to a new training (Louie and Wilson, 2001; Lee and Wilson, 2002).

The regional CBF measurements using PET carried out in human subjects have permitted to demonstrate that the least activated regions during SWS are localized in the dorsal pons, mesencephalon, cerebellum, thalamus, basal ganglia, hypothalamus, and the prefrontal cortex, which indicates that the rest of the brain remains significantly active during this stage. During PS, the especially active areas are the pontine tegmentum, some thalamic nuclei, the amygdaline complex, the hippocampus, the cingulate cortex, and the posterior temporo-occipital cortices (Maquet et al., 2005). Such promising studies are still in progress and should not lead us to fractionate the brain, as it happened in the late 19th century during the height of phrenology.

The activity of the entire CNS is necessary to generate and maintain sleep; thus, both the regions which become activated and those reducing their activity are in effect necessary to achieve a sleeping brain.

Respiratory changes

During W, respiration is controlled by a triple mechanism: (a) the metabolic, which ensures the homeostasis of arterial O_2 and CO_2 through information conveyed from the central and peripheral chemoreceptors; (b) the voluntary control, which permits to coordinate ventilation and other functions such as speech, cough, etc.; (c) the action of spinal motor neurons, which innervate the respiratory muscles and receive tonic inputs that help to maintain their membrane potential level at a certain degree of depolarization. Besides, the spinal motorneurones receive innervation from respiratory centers of the brainstem, which have semi-automatic function (Fig. 2.5). The metabolic control is integrated at the bulbo-protuberantial level and finally provokes the necessary changes to activate the respiratory muscles. The voluntary control most certainly involves structures that control brainstem and spinal respiratory neurons, e.g., the amygdala, the periaqueductal gray, frontal cortex (Orem et al., 2002; Orem, 2005).

The respiratory changes that take place during SWS and PS reflect the predominance of the metabolic control during SWS and a decrease of this control during PS. A characteristic non-homeostatic is present in PS, when ventilation no longer depends on the metabolic control.

Both drowsiness and stage II provoke an unstable respiratory rhythm with consecutive hypoventilations and hyperventilations called "periodic ventilation." During stages III–IV ventilation becomes regular, with higher amplitude and a lower respiratory rate, and a mild decrease in the per-minute volume. This is associated with a decrease in the metabolic rate and also variations in the central control of respiration. As SWS starts, automatic control mechanisms are released, with the inactivation of the telencephalic mechanisms that command during W. During stages I and II, episodes of respiratory instability alternates with periods of regular breathing, which will later stabilize in a regular breathing rhythm by stages III–IV. The partial pressure of alveolar CO_2 increases whereas the partial pressure of alveolar and arterial O_2 decreases. The chemoreceptors response to CO_2 is moderately reduced while the response to hypoxia does not change. Both, respiratory rate and depth remain relatively constant.

The respiratory rhythm during PS is typically faster and irregular, exhibiting episodes of apnea and hypoventilation. The muscle hypotonia contributes by decreasing the strength of the chest expansion and increasing the resistance to airflow in the upper air pathways. The diaphragm maintains an irregular activity and atonia of intercostal muscles occurs in PS in cats (Parmeggiani and Sabattini, 1972). While Pompeiano (1967) proposed that atonia is the result of active inhibition, both excitatory

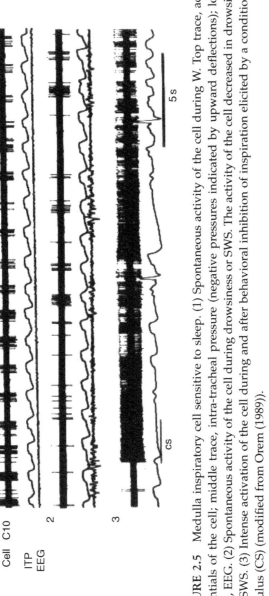

FIGURE 2.5 Medulla inspiratory cell sensitive to sleep. (1) Spontaneous activity of the cell during W. Top trace, action potentials of the cell; middle trace, intra-tracheal pressure (negative pressures indicated by upward deflections); lower trace, EEG. (2) Spontaneous activity of the cell during drowsiness or SWS. The activity of the cell decreased in drowsiness and SWS. (3) Intense activation of the cell during and after behavioral inhibition of inspiration elicited by a conditioning stimulus (CS) (modified from Orem (1989)).

and inhibitory tonic processes affect the respiratory system in PS (see review, Orem, 2005).

Endocrine functions

Blood levels of every hormone exhibit cycle. These cycles are modulated by the circadian rhythm of light and darkness, sleep and W, autonomic activity, etc., all functions which converge on the hypothalamus. The hypothalamus-hypophysis axis is the link between sleep processes and hormonal secretion.

There are many hormones that are secreted with a rhythm set by the wakefulness–sleep cycle. The influence of the sleep–wakefulness cycle on the endocrine system has been subdivided into three basic types: (a) hormones modulated by a particular stage of sleep, e.g., the peak in growth hormone (GH; Fig. 2.6) levels during a specific phase of SWS; (b) hormones highly influenced by the whole period of sleep, such as prolactin and thyrotrophin (TSH); and (c) hormones weakly modulated by sleep, such as ACTH (adrenocorticotropic hormone), cortisol and melatonin, despite exhibiting strong circadian rhythms (Brandenberger, 1993; 2005).

The GH, essential for development, shows a secretion cycle closely linked to the first episode of sleep stage IV (Takahashi et al., 1968; Sassin et al., 1969). If SWS is delayed or prevented, this hormone will not be secreted in appreciable quantities, whereas, if the SWS is facilitated, e.g., by physical exercise, this hormone will be released in high quantities. In infants and adults with prolonged SWS, secretory episodes of great magnitude can be obtained. In elderly people, coincidentally with the normal decrease of SWS, this hormone's secretion is greatly reduced. Episodes of PS take place mainly during the descending phase of the secretory pulses of GH or in its nadir.

Melatonin exhibits a secretory peak at the beginning of the night, generated by a central pacemaker located in the supra-chiasmatic

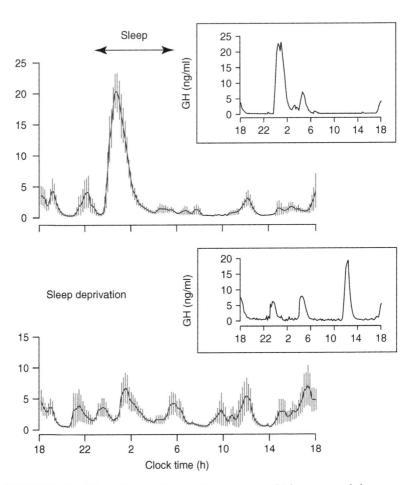

FIGURE 2.6   Effect of sleep deprivation on mean 24-hour growth hormone (GH) levels in 10 subjects. In sleep-deprived persons, the reduction of the sleep-related pulse is compensated for by the emergence of large individual pulses during the day, so that the amount of GH secreted over 24 hours is similar whether or not a person had slept during the night. Insets give a typical example of individual profiles. Bars indicate standard error of the mean (modified from Brandenberger (2005)).

nucleus of the hypothalamus, which receives projections from the retina. Its major function is the coordination of circadian rhythms. It is considered a chemical code through which the brain "understands" that it is night-time; the longer the night, the longer is the melatonin secretion, constituting in some species the temporal key to cycle seasonal rhythms (Cardinali, 2005).

Finally, all seems to indicate that the endocrine system pulsatility is commanded by complex ultradian clocks. It is important to mention that the functional significance of the diverse temporal relations between sleep and hormones is still unknown.

Physiological regulation in sleep

Physiological regulation in mammals depends on the ultradian wake–sleep cycle (Parmeggiani, 2005a). This dependency is the result of the changing functional dominance of phylogenetically different structures of the encephalon across the different behavioral states of the cycle. The functional similarity of physiological events during non-rapid eye movement (NREM) sleep in different species and the variety and variability of such events during REM sleep within and between species define the characteristic differences between these states of sleep. Intrinsic nervous processes specific to the state of REM sleep may cause somatic and autonomic variability without relationship to mental content or homeostatic control. The basic somatic features of NREM sleep are the assumption of a thermoregulatory posture and a decrease in antigravity muscle activity. The basic somatic features of REM sleep are muscle atonia, REM, and myoclonic twitches. The basic autonomic feature of NREM sleep is the functional prevalence of parasympathetic influences associated with quiescence of sympathetic activity. The basic autonomic feature of REM sleep is the great variability in sympathetic activity associated with phasic changes in tonic parasympathetic discharge. In all species, the somatic and visceral phenomena of NREM sleep are indicative of closed-loop operations automatically maintaining homeostasis at a lower level of energy expenditure compared with quiet W. In contrast, the somatic and visceral phenomena of REM sleep are characterized in all species by the greatest variability: this is a result of open-loop operations of central origin impairing the homeostasis of physiological functions (poikilostasis). The demonstration, in terms of reactive homeostasis, of different functional states of the ultradian sleep cycle that are characterized by either homeostasis (NREM sleep) or poikilostasis (REM sleep) of physiological functions is based on the criterion of short-latency stimulus–response relationships. This basic functional dichotomy applies to the nervous control of body temperature and of circulatory and respiratory functions. In contrast, gastrointestinal, endocrine, and renal functions do not fit this criterion. For example, many aspects of gastrointestinal function are not constrained within the temporal boundaries of single sleep states and appear, at most, to be modulated by changes in the autonomic nervous system outflow during sleep. On the other hand, there are changes in endocrine secretion that are specific to a single sleep state. However, such changes are the result of ultradian or circadian modulation rather than of a homeostatic response to exogenous or endogenous disturbances in terms of reactive homeostasis.

Pier Luigi Parmeggiani
Università di Bologna
Bologna, Italia

Body temperature

In homeothermic animals, the interaction between hypothalamic and cortical mechanisms controls body temperature. In SWS the automatic mechanisms are released from the cortical control. During PS, temperature regulation is interrupted. The decrease in the muscular tone and the shivering absence reduces the body's ability to produce heat. Body temperature falls along the night to reach the lowest levels by the last hours of sleep. There is a rise in skin temperature coincident with PS periods. Sleep continuity is altered by ambient conditions (Fig. 2.7; Libert and Bach, 2005).

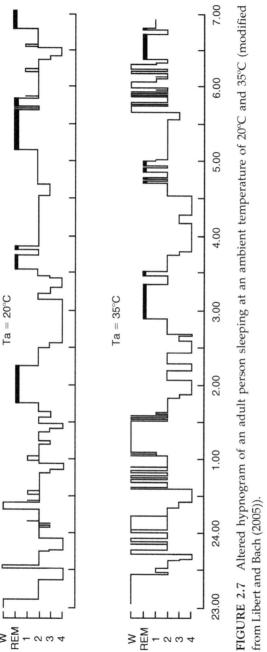

FIGURE 2.7 Altered hypnogram of an adult person sleeping at an ambient temperature of 20°C and 35°C (modified from Libert and Bach (2005)).

Works carried out by Parmeggiani (1980; 2005a) during the last decades have demonstrated that the correlation between body temperature and room temperature is variable and it depends on the moment of the wakefulness–sleep cycle being analyzed. A cat exposed to changing room temperatures is able to control and maintain its body temperature constant during W and SWS, whereas during PS its body temperature will rise or drop if its environment is warm or cold, respectively. Hence, we can assume that during PS the animal enters a state similar to that of a poikilotherm (Fig. 2.8).

Changes in other functions

Renal functions

Glomerular filtration, urine volume and sodium, potassium, and calcium excretion decrease during sleep. Urine concentration is higher during sleep than W, being the highest concentration during PS associated with an even decreased urine excretion.

Digestive functions

Studies have shown a decrease in gastric acid secretion during sleep in normal humans. Subjects suffering from duodenal ulcers have the acid secretion permanently increased over the sleep and W cycle. Recordings of bowel motility show conflictive changes up to the present time, although the esophagus motility is consistently reduced.

Sexual functions

Penile erection or tumescence has been shown in human subjects between 3 and 79 years of age during PS, although in adolescents this is not only confined to this stage. Although its functional role remains unknown, the presence or absence of

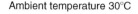

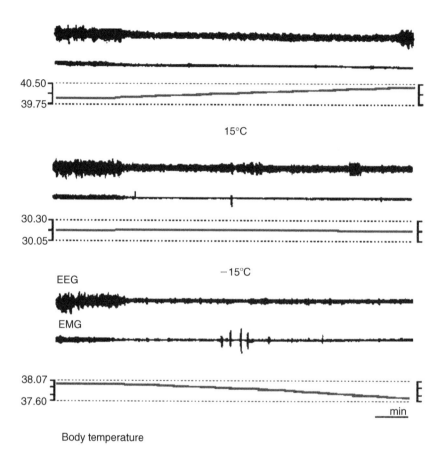

FIGURE 2.8   Body temperature shifts during PS in cats. Exposed to a high ambient temperature (30°C) the body increased its temperature. A moderate ambient temperature (15°C) is associated with stable body temperature. When the ambient temperature is very low (–15°C), the body also decreased its temperature (modified from Parmeggiani (1980)).

erection during sleep is employed for the differential diagnosis between organic and psychogenic impotence. Women exhibit clitoral erections and increment of the vaginal blood flow during PS.

Homeostasis

The existence of homeostatic mechanisms which regulate all functions was already postulated by Cannon in 1929. However, it has recently been proposed that these mechanisms may be overcharged beyond their limits during certain states, such as extremely active W. Such deviations from the homeostatic range can normally be compensated during W, restoring the functional equilibrium.

The study of sleep has demonstrated the existence of a functional disruption of the mechanisms of control, which are present during SWS and absent during PS. As a result, many of the basic functions, such as the blood pressure, breathing, temperature, etc., are left out of the strict homeostatic control and their values shift. As explained above, the body temperature becomes dependent on the room temperature, and the animal turns into a transitory poikilotherm (Parmeggiani, 1980). Furthermore, in certain regions of the cat's brain, the oxygen local availability becomes unstable during PS, which may also reflect an escape from homeostatic regulation (Velluti and Monti, 1976; Velluti et al., 1977; Velluti, 1985; 1988). We cannot yet explain the reason for the existence of a sleep stage that lacks homeostatic control. Which are the functions that require such conditions?

The energetic metabolism

It is outstanding that evolution has permitted the appearance of the energy costly and dangerously disregulated PS. Whereas the lactate produced by the glycolysis in astrocytes would be capable of ensuring the ATP production and renewal necessary for the preservation of the active W and the PS, the SWS would be a stage for saving energy. However, it is difficult to study the energetic metabolism during the wakefulness–sleep cycle due to the absence of specific pharmacological tools that permit the performance of *in vivo* studies (Franzini, 1992; Cespuglio et al., 2005).

Neurophysiology

The sleep, formerly considered a passive phenomenon, is nowadays recognized as actively generated by the brain. Experimental results such as the anatomy of lesions produced by a probable craniopharyngioma compressing the anterior hypothalamus, observed in Montevideo and published in Paris by Soca (1900), the lesions of the lethargic encephalitis published by von Economo (1930), and the demonstration of a hypnogenic thalamic region (Hess, 1944) represent pioneer studies that lend support to this concept. Sleep patterns induced by basal forebrain stimulation reinforced this notion (Clemente and Sterman, 1963).

Consistent with this concept is the demonstration that during sleep the neuronal discharge in a number of regions increases. Moreover, neuronal activity changes, increasing or decreasing, along the different stages of sleep (McGinty et al., 2005). Furthermore, a neurone that decrease its firing entering in sleep should be considered also participating of the ongoing sleep processes. Unitary recordings of motor neurones fibres and other regions show that, contrary to what occurs during general anesthesia, the discharge rate of some neurones increases or change their firing pattern during sleep, reaching even higher levels than those of quiet W (Evarts et al., 1962). Sleep increasing firing neurones were particularly observed in the preoptic region of anterior hypothalamus (Fig. 2.9; McGinty and Szymusiak, 2003).

The neuronal firing in the auditory system can increase and also decrease firing during sleep perhaps participating in some process of sleep general organization (Fig. 2.10). In Chapter 5 the auditory neurons sleep-related behaviour is shown and discussed.

Sleep, a cyclic and reversible stage, appears then as a unique physiological state of vigil consciousness abolition and reduction of response to the environment, which is accompanied by changes in multiple functions including oneiric "consciousness."

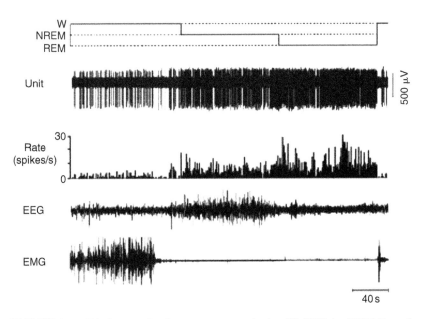

FIGURE 2.9   Discharge of a sleep-on neuron during W, SWS (or NREM), and PS (or REM), recorded in the median preoptic nucleus of an unrestrained rat. Its firing rate is low during waking, increases at sleep onset and during SWS, and reaches even higher levels in PS (modified from Suntsova et al. (2002)).

General neural approaches

Once the concept of sleep as an active process was established, the search for one or more sleep generator centers began (Kleitman, 1963). Abundant experimental evidence emerged from brain lesions, which led to the finding of regions involved in the generation and/or maintenance of sleep (Jouvet, 1961; 1962).

However, the sleep is not a function but it is instead another physiological state of the CNS and the body, which turns chimerical the search for a sleep single center.

Bremer (1935) performed the first experimental preparation called "encephale isolè," by sectioning of the CNS at the level of the first cervical segment, demonstrating that animals showed EEG and pupilar signs of W and sleep. In contrast, sectioning

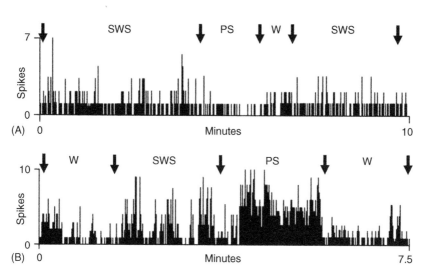

FIGURE 2.10 Unitary activity in the auditory cortex (AI) during W and sleep in a guinea pig. (A) Spontaneous discharge as a function of time. After fluctuating during SWS, the firing rate markedly decreases during PS. The number of spikes is quantified over 50 ms epochs. (B) Discharge as a function of time. The firing rate shows peaks during SWS and a *quasi*-tonic increase, on passing to PS. The number of spikes is quantified over 450 ms epochs (modified from Peña et al. (1999)).

at the level of the mesencephalon (a preparation called "cerveau isolè") produced a bioelectrical activity that resembled sustained sleep. Although intense olfactory stimulation was able to provoke brief periods of W, which did not last beyond the duration of the stimulus, visual information could not "wake up" the animal or elicit ocular signs or a diffuse EEG activation. Such experiments permitted to identify regions that participate in producing W, located in the mesencephalic reticular formation and in the posterior region of the hypothalamus. The difference between the "cerveau isolè" preparation, which exhibits alternation of sleep and W, and the large proportion of W in animals with "mediopontine" lesions indicates that a hypnogenic region responsible for EEG synchronization must be located in the posterior bulbopontine zone.

The demonstration of the existence of hypnogenic actions from structures lying in the anterior brain goes back to the pioneering experiments of Hess (1944). The electrical stimulation of thalamic regions elicits sleep in animals, which resembles the natural behaviour of the species.

By systemic injections of cholinergic drugs to pontine cats, Jouvet (1962) obtained an increase in the number of PS-like episodes. It has been demonstrated that direct application of acetylcholine crystals in the preoptic region caused sleep (Hernández-Peón et al., 1963). This effect could be blocked by applying atropine in more posterior brainstem regions, which suggested a rostrocaudal direction, along the medial forebrain bundle, of the hypnogenic influence (Velluti and Hernández-Peón, 1963). Microinjection of acetylcholine or cholinergic agents in the brainstem (Cordeau et al., 1963) and pontine region (George et al., 1964; Vivaldi et al., 1980; Baghdoyan et al., 1984; Gillin et al., 1985; Reinoso-Suárez et al., 1994) evoked PS-like signs.

W and sleep are characterized by different rhythms in the EEG and are associated with the changes in the cortical excitability that accompany behaviour. The discharge of reticular thalamic neurons synchronize the forebrain activity during SWS, inducing the slow EEG oscillations at delta frequencies and the 12–16 cycles/s frequencies typical of the spindles (Steriade, 2005).

The cerebral cortex, representing an 80% of the human CNS, must also participate in the generation and maintenance of sleep processes. Such aspects have not been sufficiently investigated so far.

Particular processes of the PS

Once in SWS, a prior and essential stage, the brain is able to produce the cortical electrographic activation and other PS signals. There are two ascending inputs to the cortex that are involved in this process: one originates in the intralaminar nuclei of the thalamus and another in the caudal hypothalamus.

Short and long PS episodes were described in rats as two populations with different functional meaning (Amici et al., 2005).

The cerebellar lesion is also capable of eliciting transitory changes in the sleep cycle. The unitary activity, both of Purkinje cells and of neurones from the fastigial nucleus, increases at the beginning of sleep suggesting a probable cooperation between this nucleus and pontine zones of common embryological origin (Mano, 1970; Hobson and McCarley, 1972). Moreover, the "pO$_2$ PS system" includes cerebellar areas also supporting the notion of cerebellar participation in PS (Velluti, 1985).

During PS the spinal motor neurons are inhibited, which explains the atonia or hypotonia of certain muscles at this stage of sleep (Pompeiano, 1967). Recent studies show that atonia may be produced by direct projections from the mesopontine tegmentum to the spinal cord (Lu et al., 2006).

Ponto-geniculo-occipital waves

The pons has a bioelectrical activity of its own that spreads throughout the brain, called Ponto-geniculo-occipital (PGO) waves. This activity appears in a very showy manner in the transition between the SWS and PS, heralding and being part of a new PS episode (Jouvet et al., 1965; McCarley et al., 1978). In the cerebellar nuclei of the cat, PGO were also reported during spontaneous normal PS, under reserpine effects, and after microinjections of acetylcholine in the pons (Fig. 2.11; Velluti et al., 1985; Velluti, 1988).

Although the various experimental approaches permitted to know the location of encephalic sites that generate *signs* of W or *signs* of the different sleep stages, it is essential to remark that are partial approximations to how the brain is functioning in order to develop a wakefulness–sleep cycle with

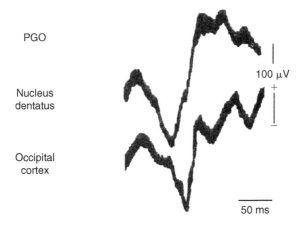

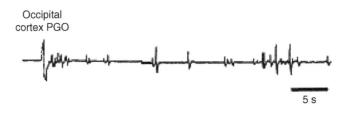

FIGURE 2.11   Example of PGO in the cerebellar nucleus *dentatus* and in the occipital cortex. PGO waves have also been found in the *fastigii* and interpositus cerebellar nuclei, with similar characteristics of amplitude and timing to PGO waves described in other areas (modified from Velluti et al. (1985)).

all its physiological characteristics. The brain as a whole is the responsible of the cycle.

A partial hypothesis on PS fragmentation

Deviations from normal values of certain vital functions such as the heart and the respiratory rates could not be indefinitely

Ponto-geniculo-occipital waves and alerting

Ponto-geniculo-occipital (PGO) waves – or spikes as they were originally named – are macropotential waves first recorded in the pontine *tegmentum*, lateral geniculate body and the occipital (visual) cortex of cats, hence the acronym. The last two sites in particular reflected the early interest in dreams and eye movements among early sleep researchers. In cats, the waves are essentially limited to about 30 seconds preceding the onset of REM sleep and for the duration of that state. Only a few waves precede REM sleep in rats. There appear to be two types of PGO waves as they are recorded in cats: those occurring singly and those appearing as bursts during an episode of REM sleep. The latter were eliminated following bilateral lesions in the vestibular nuclei that also blocked the occurrence of REM (Morrison and Pompeiano, 1966) suggesting that the PGO bursts are reinforced by the nystagmoid REM. Datta (1997) has reviewed his laboratory's work that revealed interactions between the pontine generator site identified by him and the vestibular nuclei, supporting the observations of Morrison and Pompeiano (1966). The single waves generated in the caudolateral parabrachial and subcoeruleus areas then, appear to be a primary element of REM sleep, unlike the eye movement-associated bursts.

Difficulties in recording similar waves in the lateral geniculate body of the albino rat led to some speculation that the fundamental PGO wave did not occur in this common laboratory animal, which tended to obscure what we believe to be a general phenomenon across mammalian species (Reiner and Morrison, 1980; Kaufman and Morrison, 1981). In fact, macropotential waves were later recorded in the dorsolateral pontine tegmentum that preceded the onset of REM sleep as in cats. Datta et al. (Synapse, 1998) has demonstrated that the pontine generator cells do not project directly to the lateral geniculate body in the visually deficient albino rat, possibly explaining the earlier difficulties.

The focus on eye activity and dreams may have led workers away from what is in our opinion their real significance: a sign of alerting in the brain, a fundamental – and "peculiar" – aspect of REM sleep. The idea that PGO waves are a sign of an "alert" brain largely cut off from, or at least minimally responsive to, both external and internal influences arose when Bowker and Morrison (1976) demonstrated in cats that auditory stimulation (90 dB level) would elicit PGO waves (PGO$_E$) during REM sleep and even NREM sleep. The same results were later obtained in rats.

At the cellular level, Hu et al. (1989) found that PGO waves in the lateral geniculate body of cats resulted from nicotinic activation of the projecting neurons with a parallel muscarinic inhibition of perigeniculate cells stemming from activation by pontine peribrachial cholinergic neurons. They confirmed that auditory stimuli elicited PGO waves in the lateral geniculate body and concluded, in agreement with the earlier work, that "these signals are the central correlates of orienting reactions elicited by sensory stimuli during waking (the so-called eye movement potentials) and by internally generated drives during PS."

Adrian R. Morrison
University of Pennsylvania
Philadelphia, USA

prolonged, which becomes a limiting factor for the duration of PS periods. I propose that this particular architecture of sleep is based on individuals dividing the amount of PS necessary into shorter epochs. Internal security systems would make us return to homeostatic controlled SWS or W, to enter a new PS episode after some time and thus achieving its still enigmatic and apparently essential state (Velluti, 1988).

Endogenous humoral factors

Throughout the history of sleep research, diverse factors of endocrine, biochemical origin have been postulated to explain this cycle. Pieron (1913) proposed the existence of substances which, generated during W, would be removed during the sleep. Investigations carried out in rabbits, described a non-apeptide found in the cerebrospinal fluid (CSF) after electrical stimulation of the thalamus that induced sleep (Monnier and Hösli, 1964). Because injection into the brain ventricles of other rabbits provoked delta waves in the EEG, this peptide has been denominated "delta sleep-inducing peptide." In most of the subsequent studies, this factor has only showed a mild hypnogenic effect. On the other hand, just as any other peptide, its normal passage through the hematoencephalic barrier is, in any case, difficult and slow.

Peptides derived from the propiomelanocortin and peptides immunologically active have also been proposed as hypnoinducers. More recently, two types of sleep-facilitator peptides have been proposed: (1) SWS-promoting substances (e.g., GH-releasing hormone, interleukin-1β, tumor necrosis factor α, adenosine, and prostaglandine D_2) and (2) PS-promoting substances (e.g., vasoactive intestinal polypeptide and prolactin). All these substances have fulfilled the criteria for sleep regulatory-modulatory substances (Krueger and Obal, 1994; Obal and Krueger, 2005).

Hypnogenic actions are attributed nowadays to melatonin. Recent studies have demonstrated that this hormone, known for its chronobiotic effects since a long time ago, might also have an effect on sleep, causing to decrease its latency and increase its efficiency (Cardinali and Pevet, 1998; Cardinali, 2005).

The amount of substances that have been described in recent decades permit us to conclude that none of them has a powerful and determinant action, neither can they be attributed a sleep provoking effect with physiological characteristics acting isolated. Perhaps the encompassed physiological actions of many of them generate a global modulation thus creating a situation favorable to sleep.

Neurotransmitters and neuromodulators

The study of different neurotransmitters by administration of agonists, blockers has permitted to infer that some of them play a more central role than others in the generation or maintenance of the different stages of the sleep–wakefulness cycle. Considering that neurones receive hundreds of synapses and project to varied neuronal networks, whose function changes depending on the moment of the cycle, such neurochemical studies appear as simplistic approaches. They often lead to interpret an effect as the result of the action of a certain neurotransmitter, while it may simply be the result of an imbalance in a complex neural network. Bearing in mind such limitations, we can consider certain neuronal systems and its neurotransmitters may act in certain moments of the cycle.

Among the systems involved in the generation and maintenance of W we can highlight the noradrenergic neurones in the *locus coeruleus*, the serotoninergic neurones in the dorsal *raphe* nuclei, the histaminergic neurones in the tuberomammilary nuclei, the hypocretin/orexinergic neurones in the perifornical area, the glutamatergic neurones in the reticular formation, and the cholinergic neurones in the basal brain. These systems of neurones reinforce each other and operate in specific behaviors during W by converging on common effectors in the thalamus and the cortex (Brown and McCarley, 2005).

Neurones of the anterior hypothalamus that produce gamma amino butyric acid (GABA) are involved in SWS generation. Both GABA enhancer and GABA-mimetic drugs increase the amount of SWS.

The ultradian cycle SWS–PS seems to be generated by the interaction of cholinergic and monoaminergic neurones (serotonin, noradrenalin) of the brainstem. This model indicates that when the cholinergic neurones are released from the monoamine inhibition, they stimulate the neurones in the reticular formation which in turn lead the signs of PS (see McCarley, 2004). However, selective lesions of either cholinergic or monoaminergic nuclei in the brainstem have limited effects on PS (Jones et al., 1977; Shouse and Siegel, 1992). A recent report proposed a flip-flop switch model in the regulation of the switching into and out PS (Lu et al., 2006). It was also suggested that there must be additional circuitry involved in that switching such as relevant descending pathways from the hypothalamus recently been found in rats PS (Saper et al., 2005). Lateral hypothalamic hypocretinergic neurons may act on PS-off and PS active brainstem units by means of inhibitory transmitters as galanin and GABA.

Furthermore, it should be recognized the putative relevance of the greatest anatomical part of the primates brain, the neocortical mantle. The neocortex certainly influences/determines the brainstem and other sleep-related *loci* activity, which has been, so far, almost not experimented on.

There are many neurotransmitters and hormones involved in SWS control. After diverse lesion studies and precursor injections, serotonin is nowadays considered to act indirectly on sleep induction by modulating other hypnogenic factors of the anterior hypothalamus and the suprachiasmatic nucleus (Brown and McCarley, 2005).

Acetylcholine and cholinergics

Acetylcholine microinjection techniques have demonstrated that the generation of PS-like signs is related to the activity of pontine regions (see Vivaldi et al., 1980; Baghdoyan et al., 1984; Gillin et al., 1985; Reinoso-Suárez et al., 1994). An outstanding fact is the latency of seconds or minutes between the acetylcholine injection and the beginning of the PS signs. The cholinergic substances appear to initiate a process that requires a certain time to develop or to get other regions started (Velluti et al., 1985; Velluti, 1988).

Noradrenaline might act together with acetylcholine during W, but would not operate during PS since during this stage the noradrenergic neurons of the *locus coeruleus* – at least – remain silent.

Adenosine

The purine adenosine differs from all the neurotransmitters and neuromodulators described above in several ways. It is present and released from many neurons since it is a by-product of cellular metabolism and is formed by the breakdown of the energy molecule ATP. Adenosine is not released from synaptic vesicles but instead is conveyed to the extracellular space by plasma membrane transporters. It has been implicated as a sleep-promoting factor (Porkka-Heiskanen et al., 1997; Basheer et al., 2000). In rats and cats in many areas of the brain adenosine levels rise during waking and drop substantially during SWS. In particular, in the basal forebrain adenosine levels rise during waking and importantly they continue to rise during the W induced by sleep deprivation. Moreover, adenosine inhibits identified cholinergic neurons *in vitro* whereas *in vivo*, adenosine inhibits wake-active neurones (Brown and McCarley, 2005).

Recent studies have found the hypocretin/orexin neuronal neuromodulator system participating in the motor activity of both W and PS. This system would also be involved in certain pathologies related to sleep, such as the cataplexy of the narcolepsy (Taheri et al., 2002; Mignot, 2004).

Possible functions performed during sleep

We know that the body functions are influenced by the alternation of W and sleep; however, it is still unknown why we sleep. The most accepted hypotheses at present about the biological functions of sleep can be summarized as follows.

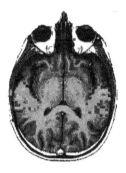

PLATE 1.11   Auditory functional MR imaging with music. The intensities of brain regions activation varied from high (yellow) to low (blue). The primary auditory cortex is bilaterally activated while the activity of secondary auditory cortex is greater on the left (modified from Bernal and Altman (2001)).

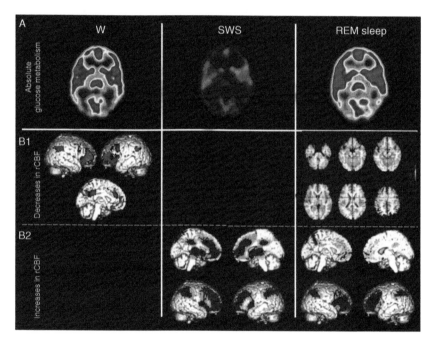

PLATE 2.4   Glucose metabolism and regional CBF during W, SWS, and PS (REM). (A) Cerebral glucose metabolism quantified in the same individual at 1-week intervals, using fluorodeoxyglucose and PET. There is a significant decrease in the average glucose metabolism during SWS compared to W. During PS (REM) the glucose metabolism is as high as during W (Maquet et al., 1990). (B1) Distribution of the highest brain activity, assessed by CBF measurement using PET during W and PS (REM sleep). The most active regions during W are located in the associative cortices in the prefrontal and parietal lobes (Maquet, 2000). During PS (REM), the most active areas are located in the pontine tegmentum, the thalamus, the amygdaloid complexes, and the anterior cingulate cortex (Maquet et al., 1997). (B2) Distribution of the lowest regional brain activity during SWS and PS (REM sleep) using the same method as in B1. In both sleep stages, the least active regions during W are located in the associative cortices in the prefrontal and parietal lobes. During SWS, the brainstem and thalamus are particularly deactivated (modified from Maquet et al. (2005)).

Tone-burst stimulation during sleep

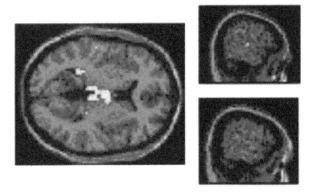

Name stimulation during sleep

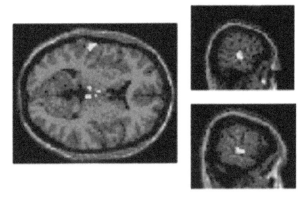

PLATE 4.7   Brain areas activation both during auditory stimulation with a non-significant sound, tone-bursts, and with a significant one, the person's name, during sleep and in comparison to W (modified from Portas (2005)).

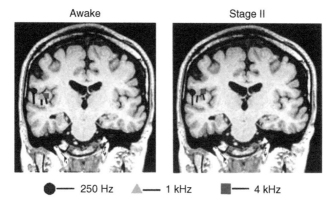

PLATE 5.17 *Planum temporale* auditory cortical location of the M100 MEG component observed in response to three different sound frequency stimuli (250 Hz, 1 and 4 kHz) recorded in W and stage II sleep in human. The magnetometer was placed on the left hemisphere (C3 position) and the signal source was estimated using an equivalent current dipole (ECD) model. ECD overlapped on magnetic resonance imaging. The ECDs (dipoles) were localized deeper in response to the higher-frequency tones to the lower-frequency tones, while the three changed position on the cortex. The relatively great shifts in the cortical space exhibited by the dipoles demonstrate that the working network changed, surely including new cells elements and communications (modified from Naka et al. (1999)).

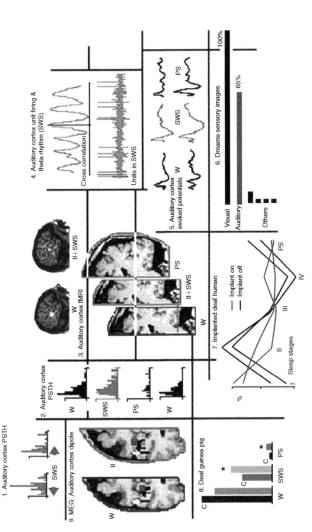

PLATE 6.4 Diverse technical approaches supporting the postulated notion of the importance and possible active participation of the auditory input on sleep processes. Three human half-brain tomographic cuts (center) represent the three main functional possibilities: W, wakefulness; SWS, slow wave sleep; and PS, paradoxical sleep. (1) Post-stimulus time histogram (PSTH) changes of a cortical auditory neurone firing shift when stimulated with natural sound played directly or backward (Pérez-Perera et al., 2001). (2) PSTH of a cortical unit on passing from wakefulness to SWS and PS exhibits firing and pattern shifts (Peña et al., 1999). (3) Human auditory cortical imaging (fMRI) demonstrates activity during sleep (modified from Portas et al. (2000)). (4) The cortical auditory neurones can be phase-locked to hippocampal theta rhythm (Pedemonte et al., 2001). (5) Rat auditory cortical evoked potentials through the sleep–waking cycle show amplitude changes (Hall and Borbély, 1970). (6) The dream auditory "images" are present in 65% of dream recalls (McCarley and Hoffman, 1981). Human and guinea pig deafness influence sleep: (7) the human recorded with the intra-cochlear implant off and on shows different sleep stages percentages while (8) the guinea pig exhibits (bars) an increase in sleep time with decreasing wakefulness (Pedemonte et al., 1996b; Velluti et al. 2003). (9) The human MEG shows a place shift of the dipole evoked by three sound stimulating frequencies on passing to sleep stage II, demonstrating a change of neuronal network/cell assembly (modified from Kakigi et al. (2003)).

Recovery and restoration

The hypothesis that the sleep serves the recovery and restoration of biochemical and physiological processes degraded during W is apparently logical and widely accepted. The increase in GH during SWS in humans would support this idea, although there are other species that do not exhibit such temporal correlation (e.g., *rhesus* monkeys and dogs). Whereas in humans there is a correlation between the duration of the preceding W and the subsequent sleep, this fact is not so clear in other species.

The effect of physical exercise on the subsequent sleep does not support the hypothesis of the restitution of the body in general. Contrary to this hypothesis is the fact that exercise carried out during the hours prior to sleep provokes delays in its installation and a lag of the circadian rhythm.

The CBF during SWS is always lower than during the corresponding PS, but the absolute value of SWS CBF "at the beginning of the night can be higher than the PS value at the end of the night. Thus, reduced SWS (0.25–4.0 cycles/s) and reduced metabolic activity towards the end of sleep suggest that some kind of recovery has occurred" (Zóccoli et al., 2005). Furthermore, Tononi and Cirelli (2003; 2005) have proposed the synaptic homeostatic hypothesis. While wakefulness is associated with sysnaptic potentiation in many cortical networks, slow wave activity is associated with sysnaptic downscaling. Sleep would counteract the increase in synaptic strength occurring during wakefulness.

The functional benefit would be to diminish the energy requirements of the cerebral cortex during slow wave sleep allowing a kind of recovery.

Energy conservation

During SWS inactivity, the lower metabolic rate and body temperature reduces energy consumption. The metabolism reduction during sleep is of about 10% in relation to the basal levels of W. However, the energy conservation can be higher at low temperatures. The metabolic reduction in an unclothed human subject exposed to a room temperature of 21° may reach a 40% after sleep onset.

Plasticity, memory, and learning

Sleep appears essential for cognitive development, although we are still far from knowing what processes are triggered during this period. However, memory and learning appear as relevant functions affected by sleep.

Recent data show that the learning process improves when followed by a night's sleep; it could be inferred that the SWS would be more closely related to the brain plasticity than to the total organism restoration (Cipolli, 2005). In addition, memory loss, learning difficulties, decrease in motor skills, and mood changes are among the earliest signs of sleep deprivation.

Two general hypothesis of how sleep participates in memory consolidation have been proposed: (1) the *dual-process hypothesis* conceives that procedural memory benefits mainly from PS and that declarative memory is mainly consolidated during SWS II–IV stages. (2) The *sequential hypothesis* assumes that the occurrence of both SWS and PS is necessary to improve consolidation of adaptive and especially of procedural memories (Giuditta et al., 1995; Stickgold et al., 2000a,b; Cipolli, 2005). Moreover, an explanatory hypothesis has been put forward, namely that the consolidation of memories acquired in a previous waking is due to the off-line reprocessing that they undergo during sleep (Stickgold, 1998).

The mismatch negativity (MMN) is an electrophysiological signal that can be correlated with learning, which changes along PS (Atienza et al., 2002). Furthermore, if a breast-fed infant is stimulated with a regular series of auditory stimuli and deviant stimuli randomly alternating during SWS – the deviant stimuli producing an MMN – this response is repeated with the same auditory series during the following W (Cheour et al., 2002). Thus, newborns can assimilate auditory information during sleep. Furthermore, animal research shows that neurones recorded from the primary auditory cortex of a guinea pig during SWS can discriminate a natural stimulus – conspecific vocalization – from its copy inverted in time (see Chapter 5). All of this makes the sleep as a whole essential for a normal cognitive development and organization of the information obtained during W.

As a first general conclusion, it seems evident that the main objectives of the SWS and the PS remain unknown. It has been demonstrated that the ultimate general objective of sleep is not to provide a resting period to the CNS or the body. Moreover, although today it is not possible to establish the ultimate reason of sleep, according to what we have seen, they must be highly diverse and certainly cannot be postponed.

Finally, I will introduce the idea of *"psycho-physiological information homeostasis"*. The learning processes, particularly those carried out during sleep, would decide what data to remember or forget in a moment of life and/or throughout life. To remember requires to select a memory trace within irrelevant memories, thus there is cost of selecting target memories (Kuhl et al., 2007). The *psycho-physiological information homeostasis* maintains the memory load within a range that the brain needs to maintain in order to preserve the cognitive functions. To remember of to forget are part of our brain goals in order to embrace a healthy status.

CHAPTER

3

Notes on information processing

In general terms, information processing can be defined as a shift of information in some manner detectable by an observer or a system. In relation to biological sensory systems, it may be a process which describes all that happens in the outer world or in the body, e.g., nature sudden sounds, the changing of the heart rate when rapidly awake or after running, a headache, and so on.

This processing has been more specifically defined by Shannon (1948) as the conversion of latent information into manifest information: "The fundamental problem of communication is that of reproducing at one point, either exactly or approximately, a message selected at another point" (Fig. 3.1). From a cognitive approach, information processing is an instrument to try to reach the understanding of human memory storage or learning processes. Therefore, equivalent objectives can be followed also in sleep.

The information processing can be sequential or in parallel, both of which can be either centralized or distributed. The parallel distributed processing has lately become known under the name of connectionism. The idea of spontaneous order in

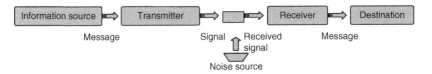

FIGURE 3.1   Communication system according to Shannon. The information source selects a message and the transmitter changes the message into a set of signals, which are in turn sent to the receiver. The receiver converts back the signal into the message. Noise is added to the signal (modified from Shannon (1948)).

the brain arising out of decentralized networks of simple units, neurones, was already put forward in the 1950s.

The world is a highly structured place. This structure is reflected by the fact that signals that reach our sense organs are not completely at random, but rather exhibit correlations in space and time. The receptors, pathway nuclei, and cortical *loci* process and distribute the space and time information; thus, the brain must deal with both place neurones – space-related units represented in the hippocampus – and our theta rhythm phase-locked auditory units, i.e., time correlated (see Chapter 5).

Information entropy is the average number of bits needed for data storage or communication. Experiments in several systems demonstrate that real neurones and synapses approach the limits to information transmission set by the spike train or synaptic vesicle entropy. Rather than throwing away information in favor of specific biologically relevant signals, these neurones seem to pack as much information as possible into the spike sequences they send to the brain, so that the same number of spikes can be used to transmit more information about the most structured signals coming from the real world (Rieke et al., 1997).

Coding

For any kind of neural processing, and particularly for auditory data processing, a basic code is needed. In general, it

Cell assemblies and neural networks

The concept of cell assembly has been one of the bases for the development of mathematical models of distributed memories, an enterprise performed by various researchers around 1970. These models showed how information encoded in thousands of parallel firing neurones can be stored into a large neuronal network. The mathematical representation of the global activity of a large cell assembly is a large-dimensional numerical vector. This vector is composed by numbers that measure the biophysical activities of the neurones (e.g., firing rates). Some memory models store pairs of input–output vectors. The material counterpart of these models suggests that the physical residences of memory traces are the synapses (Arbib, 2002).

A consequence of these models was the creation of many biologically inspired learning algorithms. These algorithms are mathematical procedures that usually modify the synaptic strengths in response to the intensity of the local synaptic inputs and the global neuronal output. Among the powerful learning rules, let us mention the Widrow–Hoff algorithm for single neuronal layer memories, and the Backpropagation algorithm for multilayer memories (or memories with "hidden layers"). The distributed memories models have interesting "biomimetic" properties. For instance, in large networks partial damages of the physical support of the memories (e.g., destructions of neurones or synapses) do not produce catastrophic deteriorations of the stored data. Other example of biomimesis is the following: some distributed memories models are capable of building prototypes from a partial exposition to data, a property that emulates the cognitive construction of general concepts from partial experiences; these prototypes are statistical averages emerged from the experienced perceptual inputs and are represented by large-dimensional vectors.

The neural representation of the human language is an ambitious (and yet far) objective of the neural network theory. Nevertheless, some important advances have occurred in recent years. In this sense, we mention the model proposed by Elman to represent the language processing as a dynamical time-depending activity. This model uses a set of modular neural networks that performs the conversion of phonetic inputs into conceptual outputs, and includes an associative memory and a working memory. The working memory produces contexts that help to decode the phonetic inputs entering the conceptual memory during the subsequent time steps. The original Elman's model uses associative memories containing hidden layers, and a miniature language is installed in the memory using the backpropagation learning algorithm. Recently, this model has been adapted by Hoffman and McGlashan to explore aspects of the neurodevelopmental hypothesis about the origin of schizophrenic disorders. In particular, these authors focused on the emergence of verbal hallucinations, and created a computational protocol to numerically simulate the generation of hallucinated voices in neural networks with exacerbated synaptic disconnections. Hoffman–McGlashan approach can be expanded in various directions, and the approach can be implemented using different neural models. All these models allow to experiment "in numero" the effect of different pharmacological (together with psychological) therapies on the hallucinations. Neural networks models offer to medical research a promising integrative instrument to help with experimental and clinical research (Spitzer, 1999).

Eduardo Mizraji
Universidad de la República
Montevideo, Uruguay

is likely that this basic code is that of a cell assembly, i.e., the assembled activity of a population of neurones (Sakurai, 1999). Understanding the neural coding signifies to understand the relation between the events in the sensory incoming data from the world or body and the spike trains. We may assume that the code is similar during the dissimilar brain states considered, i.e., wakefulness and sleep stages. The differences between states would then be based on the configuration of the new organized neuronal network/cell assembly active regions. The final result of the processing during wakefulness or sleep will be different taking into account that the brain regions involved have changed. That is, the brain is, in a large scale, functionally different upon passing to sleep phases.

A dynamic modulation of the correlated neuronal firing may occur, i.e., an auditory neurone that participates in a certain functional cell assembly/network may later on become associated with another activated and perhaps competing neuronal assembly on passing from wakefulness into sleep. The partial overlapping of neurones among assemblies is due to the ability of one neurone to participate in different types of information processings, based on the enormous amount of synapses each neurone may receive. Moreover, this condition may be repeated during the many and diverse waking states, during human sleep stage I or II and slow wave sleep (SWS), and also during phasic or tonic epochs of paradoxical sleep (PS).

It is usually assumed that information is carried by the firing rate. Modulation of many biological sounds (frog call, guinea pig call, speech, etc.) occurs in a very short time, 5–20 ms. During this time, a neurone that normally fires, e.g., 100 spikes per second, can generate just one or two spikes. A "timing code" includes both the spike count and the time quantity measuring the probability of spike occurrence. Barn owls can localize a prey by acoustic cues alone, with an auditory system that can recognize different times of signal – spikes – arrival as low as 1 ms. There is no question that this temporal information is essential to such discrimination task. It has been identified that the neural circuit is responsible for the precise temporal measurement of phase-locked spikes coming from the two ears

(Carr and Konishi, 1990). Moreover, different temporal spike patterns in the presynaptic nerve fibres may result in a completely different postsynaptic response. Two different patterns of presynaptic experimental pulses demonstrated that the presynaptic timing is important in determining the postsynaptic response (Segundo et al., 1963).

Neuronal network/cell assembly

The concept of neuronal assemblies is defined by the temporally correlated neuronal firing associated with some functional aim and the most likely information coding is the ensemble coding by cell assemblies. Neuronal groups connected with several other neurones or groups can carry out functional cooperation and integration among widely distributed cells even with different functional properties to subserve a new state or condition. On the other hand, an individual neurone receives thousands of synaptic contacts on its membrane that turn its activity into a continuous membrane potential fluctuation, which determines a very instable physiological condition to constitute a basic code for information processing. Furthermore, the neuronal network/cell assembly may provide a selective synaptic activity enhancement referring to a dynamic and transient efficacy which I suggest to be correlated to the behavioral dynamic modulation of the sleep process. That is, a neurone firing in a functional associated group may process some information and, some time later may become associated with other competing and activated neuronal groups for different functional purposes, e.g., after passing from wakefulness into sleep.

These diverse and new neuronal associations may occur during the wakefulness states, during SWS (III–IV), in human sleep stage I or II and also during PS, phasic or tonic epochs.

Figure 3.2 explains the basic possibilities or properties of a cell assembly coding. Schematically, it shows a partial overlapping

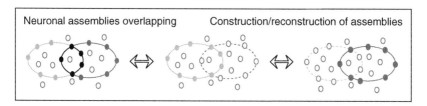

FIGURE 3.2   Examples of the manifold possible encoding properties and neuronal network/cell assembly combinations of active neurones. The arrows indicate possible and minimal dynamics of constructions and reconstructions of cell assemblies. This is an oversimplified approach of what can occur throughout the brain during the sleep–waking shift.

of neurones. Some of them belong to two different neuronal networks while a second physiological possibility is the switch of others from one state to another, i.e., construction and reconstruction of assemblies.

Progressing with this analysis, let us compare what happens in the anatomo-functional "brain" networks of a simple mollusc, *Tritonia*, capable of two different motor abilities (Fig. 3.3) which, extrapolating, could represent two different basic components in a complex brain such as the waking and sleep states. Getting (1989) postulated that "… If these network, synaptic, and cellular mechanisms are under modulatory control, then an anatomical network may be configured into any one of several modes … The term modes is intended to imply a manner in which a network processes information or generates an output pattern… ." When afferent or modulating inputs alter the properties of the basic constituents of a set of networks, a transition among modes may occur, e.g., in the present case passing from wakefulness to sleep. A neurone, as a basic constituent of a network or cell assembly, may fire an action potential or not, or may increase or decrease firing while still belonging to the same network although participating in a new particular function. However, increasing firing does not necessarily mean that a cell is subserving two different processes such as sleep

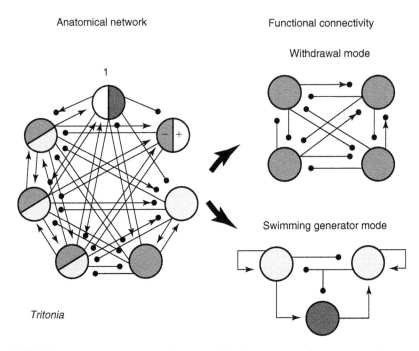

Anatomical network

Functional connectivity

Withdrawal mode

Swimming generator mode

Tritonia

FIGURE 3.3   Neuronal network of the *Tritonia*, anatomical monosynaptic connectivity – excitatory and inhibitory – and the two possibilities for functional–behavioral networks acting as: (a) withdrawal mode and (b) swimming generator mode. The neurone 1 (top, left) activation–deactivation is the first action to produce the functional and, therefore, anatomical reorganization shift, changing the animal behavior (modified from Getting (1989)).

or waking. While sleep takes place, case for instance, a recorded neurone may belong to a network in which, although engaged in sleep, might play a decreasing firing *mode*.

The difference introduced by cell 1 in the *Tritonia* movements by functionally constructing new networks (Fig. 3.3) could be produced, in a complex brain, by a summation of factors. After some unknown signal(s), e.g., decreasing light intensity, a group of neuronal networks/cell assemblies progressively begin to change the system into a *sleeping mode*. This assumption is partially supported by the observation that when a human or animal is entering sleep, the many variables usually recorded never occur in synchrony, instead appear with seconds of difference among them (electroencephalography slow activity, electromyogram decrement, eye movements, hippocampal theta rhythm frequency and amplitude, heart rate shifts, blood pressure changes, respiration rate alterations, and so on).

Sleep, on the other hand, is not a function but a complete different central nervous system state. Hence, sleep means a whole change of networks/cell assemblies, a new cooperative interaction among them, considering that a single network may subserve different functions when asleep or awake. Besides the auditory unit firing in guinea pig waking and sleep (Chapter 5), there is another available demonstration of the neuronal networks subserving auditory changes on passing to sleep, which is shown in Figure 5.17: the auditory evoked magnetoencephalography dipole location shift on the *planum temporale*, indicative of the fact that other set of neurones became engaged in the auditory signal processing in human sleep stage II.

New sets of neuronal networks/cell assemblies are reorganized on passing from waking to sleep, therefore constituting a different state, while they continue to be receptive to sensory incoming information.

4

Auditory information processing during sleep

The sensory incoming information is a relevant factor to be considered when waking and/or sleep are under question. The following experimental data will introduce the subject leading to further analysis of the relationships between audition and sleep.

Without sensory input

The surgical section of the olfactory, optic, statoacoustic, and trigeminal nerves, one vagus nerve, and the spinal cord posterior paths in cats, a *quasi*-total deafferentation, carried out by Vital-Durand and Michel (1971), reconfirmed a decrement in motor activity as previously reported by Galkine (1933) and Hagamen (1959) with a predominant sphinx head and body position. Thus, the decreased contact with the external world may induce particular behaviors and may also result in the lack of behavioral initiatives. The former data once cited led to the erroneous idea of a "continuous sleep," 99% of the time according to Hagamen (1959), in an almost totally sensory-deprived cat. When studied with polygraphic controls (Vital-Durand and Michel, 1971), the animals under *quasi*-total deafferentation condition revealed a sleep–waking cycle showing different characteristics:

- The waking time was reduced from 44.9% to 18.5% and when asleep the cats could be awakened easily at any moment. The electroencephalography (EEG) characteristics were normal with the exception of slow wave bursts in the visual cortex (Kasamatsu et al., 1967).
- The time spent in slow wave sleep (SWS) was reduced from 41.7% to 29.6%. An almost constant "somnolence" was described associated with the sphinx position and a sequential fast and slow EEG activity. On the other hand, the hippocampus and amygdala activity was that of a quiet wakefulness (W) indicative of a distinct state, both from a behavioral and from a bioelectrical viewpoint.
- The total amount of paradoxical sleep (PS) was slightly diminished (from 13.4% to 11.2%) with normal episodes length and frequency. PS showed the characteristic phasic signs such as eye movements, middle-ear muscles and muscular twitches, present prior to the cervical spinal cord lesion. Muscle activity inhibition as well as cortical EEG activation, hippocampal theta rhythm, and ponto-geniculo-occipital (PGO) waves in the visual cortex were also present. Moreover, the sensorimotor cortex has an active, desynchronized EEG and the hippocampus exhibits a normal theta rhythm; the visual cortex showed short periods of flat EEG between bursts of high-amplitude fast activity.

After sensory deprivation in cats, a new behavioral state seems to develop. Considering the behavior and the bioelectrical activities recorded, a new state could be described under what was called *somnolence*. The cat sphinx behavior observed was associated with a peculiar occipital EEG and normal sub-cortical activity, thus characterizing what I am postulating as a new although abnormal brain state dependent on the lack of sensory input.

Total auditory deprivation in guinea pigs by surgical removal of both cochleae enhances SWS and PS in similar proportion, while reducing W (Pedemonte et al., 1996b; Cutrera et al., 2000). The SWS and PS increments cited above were determined mainly by an increase in the number of episodes, but there was no change in the duration of a single episode. Authors assert that the relative isolation from the outside world may be in part because of the change observed in deaf guinea pigs. Thus, the elimination of an input to a complex set of networks, such as the ones that regulate the sleep–waking cycle, would introduce functional shifts particularly if such input has some significance, as appears to happen in the case of the behavior under study: W and sleep.

Human sensory deprivation experiments are different since they should be better considered as a reduction of sensory input, as much as possible, which leads to the notion that when a human subject is placed in an environment without patterned and changing stimulation, they may fall into a state of lowered arousal and sleep (Zubek, 1969).

Evoked activity

Two types of evoked responses can be recorded in humans: near-field and far-field evoked potentials. In animal experimentation, a local-field activity can be recorded with electrodes placed into the nuclei or cortex. The evoked potential recorded from the scalp represents a kind of mixture of both, near-field and far-field potentials. The evoked potentials provide information about the sensory processing, to localize a lesion site, to add data about the maturation or of an aged brain. Moreover, two types of response are usually distinguished in humans: *sensory* and *cognitive* event-related potentials.

During sleep, a normal reaction to any supra-threshold sensory stimulation is a return to an awake condition. Moreover, upon sensory stimulation, a special evoked EEG pattern was described in SWS (Loomis et al., 1938) designated as the K-complex. Apart from appearing spontaneously, K-complexes were observed in response to sensory stimulation such as visual, somesthetic, and auditory, being the latter the most effective (Halász and Ujszászi, 1988; Bastien et al., 2002). The K-complex response to auditory stimulation was large and less variable during stage II sleep, while during SWS (III–IV) there were no sensory-evoked modifications of the electrical activity (Davis et al., 1939). Spontaneous K-complex may be due to interoceptive stimuli and some components of the evoked K-complexes show habituation in response to repetitive stimulation (Bastuji and García-Larrea, 2005).

Sensory evoked responses

The sensory data input and data integration are not abolished during sleep. Several electrophysiological reports – from human and animal experimentation – as well as the easy awakening with significant sounds and the incorporation of stimuli into dreams, are results indicative of the conservation of some kind of processing in sleep, surely different from the one observed during W.

Human auditory responses recorded from vertex have been reported by several investigators using similar approaches and obtaining similar results (Fig. 4.1). In all subjects the auditory evoked responses exhibit major changes on passing from the wake state to the stages I, II and SWS, and a consistent increase in peak-to-peak amplitude was present while, during PS, the amplitude was decreased, approximately to that of the wake state. The latter waves of the response were of longer latency during both sleep phases, SWS and PS (Vanzulli et al., 1961; García-Austt,

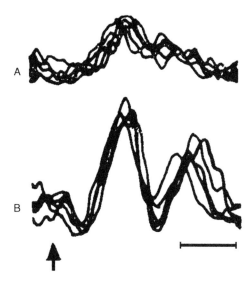

FIGURE 4.1   Auditory evoked potentials during W (A) and sleep (B) in human. Superposition of five averaged responses showing amplitude increment and waves complexity on passing to sleep. A mixture of both near-field and far-field potentials are part of this response. Click stimuli at 1/s (modified from Vanzulli et al. (1961)).

1963; Davies and Yoshie, 1963; Williams et al., 1963; Weitzman and Kremen, 1965; Ornitz et al., 1967).

Human auditory far-field evoked potentials

In a comprehensive review of the evoked potentials and information processing during sleep, some interesting conclusions were drawn (Campbell et al., 1992). The experimental data gathered using the far-field potential recording technique in humans showed no sleep effects on the brainstem auditory evoked potentials (BAEPs) (Amadeo and Shagass, 1973; Picton et al., 1974; Osterhammel et al., 1985; Bastuji et al., 1988).

However, Bastuji et al. (1990) have reported small BAEPs latency changes during nocturnal sleep. In addition, the constancy of the response was maintained whether sound stimuli were either of high or low intensity (Campbell and Bartoli, 1986).

The BAEP are far-field potentials technically recorded distant from the brainstem nuclei, being a coarse representation of auditory function that do not reflect the effects of sleep. On the other hand, the auditory nerve compound action potential (cAP) in sleep, described in guinea pigs (Velluti et al., 1989), or the effects of sleep on the guinea pig auditory nuclei unitary firing have shown significant cAP amplitude and unit firing shifts (see review, Velluti, 1997; 2005; Pedemonte and Velluti, 2005a). The two technical approaches, human and animal, have lead to different results that deserve more research for the reason that, using a more sensitive technique, sleep related changes will also appear in humans.

Another phenomenon also points out sleep actions on the auditory receptor itself, namely the transiently evoked otoacoustic emissions (sounds emitted by the cochlea reflecting the cochlear outer hair cell motility controlled by the efferent system, i.e., the CNS). It has been reported in humans as being modified by sleep in general although independently of the sleep phase (Froehlich et al., 1993).

Data regarding the sleep effects on middle latency auditory evoked potentials (potentials perhaps arising from the reticular formation, thalamus, and primary cortex) are no consistent. While early studies indicated that these components were either not affected or only slightly affected by sleep, more recent reports showed marked changes most notably on the later evoked potential components (Mendel and Goldstein, 1971; Osterhammel et al., 1985; Erwin and Buchwald, 1986; Ujszászi and Halász, 1986; Deiber et al., 1989), although Campbell et al. (1992) suggested that these waves were attenuated only with fast stimulation.

Long latency evoked potentials

The later components of the evoked potential, also called the slow potentials or late auditory evoked responses, are most altered during sleep. Long latency W auditory evoked potentials are biphasic, negative–positive complexes, N_1–P_2, at 100–150 ms post-stimulus. The neural generators of the N_1–P_2 responses are not well known. However, both modeling as well as intracranial recordings have located the sources in secondary auditory cortical areas.

Sleep is characterized by a latency delay and amplitude decrease of the N_1, associated with an enhancement of P_2 (see Campbell, 1992). Shifts in N_1–P_2 during sleep onset are fast and time-locked to sleep entrance which has been proposed as a marker of sleep onset (Ogilvie et al., 1991; Campbell, 1992; de Lugt et al., 1996). After sleep entrance, the N_1 and P_2 modifications persist with little change during both SWS and PS (see Bastuji and Garcia-Larrea, 1999; 2005).

Cognitive, event-related potentials

Event-related potentials (ERPs) can be recorded by introducing deviant, unexpected stimuli within a stimulus train. The ERPs may be part of cognitive processes: capacity of discrimination, attention, memory, and so on. High-level sensory integration, using equal stimulating paradigms, have validated the possibility of such processing continuity during the full sleep–waking cycle.

P300

The P300, a human bioelectrical component, appears in association with deviant stimuli delivered at random in a stimulating regular train, i.e., the discrimination of an odd-ball stimulus. P300, a positive component with the highest amplitude between 220 ms and 350 ms, is usually associated with attention and discrimination. The possible P300 generators are the prefrontal cortex, the temporal and parietal cortices, the hippocampus, and the cingulate gyrus, all contributing to its generation (Baudena et al., 1995; Brázdil et al., 1999; 2001; Escera et al., 2000). The P300 latency increases and amplitude decreases from W through sleep (Wesensten and Badia, 1988; Nielsen-Bohlman et al., 1991; Harsh et al., 1994; Atienza et al., 2001; Cote, 2002). Apparently, P300 can be recorded during the transition from W to stage I sleep, reappearing during PS. The lack of the frontal component of P300 may support the notion of semi-automatic, non-conscious detection of stimuli during sleep.

The averaged ERPs (P300) to regular and deviant tones during W, stage I, and PS are shown in Figure 4.2 exhibiting shifts on passing to sleep. Moreover, although with some differences (not shown), changes are also present in stage II and SWS.

The results exhibited are indicatives of a capability of the sleeping brain to detect equally physical stimuli among a group of regular ones thus reacting to novelties. These facts do not necessarily imply that their underlying processes are equivalent to those of occurring in W.

Mismatch negativity

The mismatch negativity (MMN) is elicited by deviant stimuli occurring at random within a stream of regular tones. It is acknowledged as reflecting an automatic detection of a series of new sensory inputs that does not "match" the neuronal representation of the regular stimuli, an automatic comparison process (Näätänen and Alho, 1995). The MMN is considered to reflect non-conscious activity of the sensory processing, related to the temporal and frontal networks activation (Fig. 4.3). The

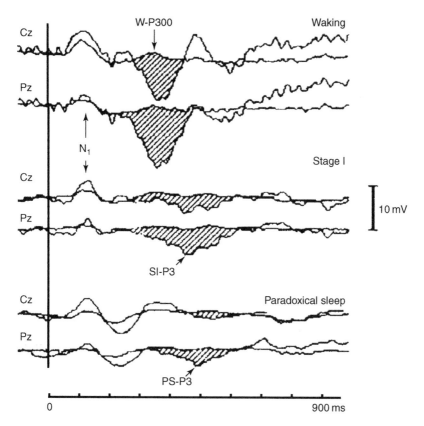

FIGURE 4.2   ERPs, P300, obtained with pure-tones in a classical odd-ball paradigm. P300 in one subject during waking, sleep stage I, and PS (modified from Bastuji et al. (1995)).

analysis of this response in sleep is a step forward in studying the sensory processing. While the persistence of a system of deviance detection during stages I, II and SWS is supported by a bulk of convergent studies, the available evidence suggests that such detection does not involve the generation of a cerebral response comparable to the waking MMN (Atienza et al., 2002).

Contrary to results in stages I, II and SWS (Campbell et al., 1992), other investigators have reported a genuine MMN

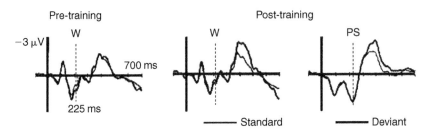

FIGURE 4.3   MMN. Grand average waveforms to standard (thick line) and deviant (thin line) stimuli during W pre-training and after training, and post-training during PS (modified from Atienza and Cantero (2001)).

when recording the stage II and during PS (Loewy et al., 1996; Atienza et al., 1997). The PS-MMN appeared to be of similar latency but smaller amplitude relative to its waking counterpart. Interestingly, this component has been reported during quiet sleep (equivalent to adult SWS) in newborns (Cheour et al., 2002).

However, a possible influence of technical factors hampering the recordings, e.g., unfavorable signal-to-noise ratio relative to waking, should not be dismissed, since this very small component may be lost within the much higher sleep negativities. However, the auditory cortex can maintain during sleep the ability to organize the arrived information although with great differences with the corresponding one in W. The MMN system activates memory when it does not match related immediate information (Atienza et al., 2002). Moreover, the well-known bottom–up influences (Velluti and Pedemonte, 2002) are added to the sleep top–down actions, particularly shown in the auditory system through its efferent fibres system (Velluti, 1997; 2005).

Experimental animal studies of evoked potentials

The amplitude of the averaged auditory nerve cAP and the microphonic potentials, in response to clicks and tone-bursts,

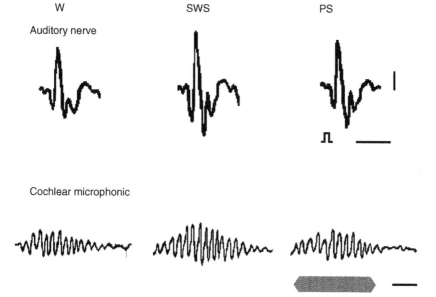

FIGURE 4.4   Auditory nerve cAP and cochlear microphonic (CM) during W, SWS, and PS in the guinea pig. Both signals were recorded from the round window with an implanted macro-electrode. The CM was evoked by a pure tone-burst (1 kHz) and the cAP by a click (0.15 ms). CM and cAP were averaged ($n=30$) and in both cases the amplitude increased during SWS and decreased during PS in comparison with a quiet waking period. Calibration bars are: cAP, 5 ms, 50 μV; CM, 5 ms, 200 μV (modified from Velluti et al. (1989)).

has been reported to change throughout the sleep–waking cycle of the guinea pig (Velluti et al., 1989). Both potentials recorded in response to low-intensity sound stimulation during sleep phases changed in a roughly parallel fashion, increasing amplitude during SWS and decreasing during PS to amplitudes similar to those observed in W (Fig. 4.4). These results were more recently repeated using new devices, applying new techniques while still using guinea pigs (Pedemonte et al., 2004).

This modulation of the auditory peripheral potentials by the sleeping brain – through activation of the efferent terminals on the micromechanical properties of the coupled efferent bundle-hair cells and direct actions on the afferent fibers – results in, among other things, a modulation/gating of the auditory input ensuing in a facilitation during SWS, a reduction of the input during PS and variable amplitudes during W. A corollary to this statement is that the changes observed in the peripherally recorded auditory potentials during W and sleep phases may condition, at least in part, the described shifts in the responses evoked upstream in the pathway, as far as the auditory cortex itself.

The CNS auditory evoked potentials changes associated with sleep, W, or anesthetics have been reported in cats by several authors (Huttenlocher, 1960; Jouvet, 1962; Pradham and Galambos, 1963; Teas and Kiang, 1964; Herz, 1965; Herz et al., 1967; Petrek et al., 1968; Wickelgren, 1968) and in rats (Hall and Borbély, 1970). Several discrepancies among the reports by different authors were observed in cats as well as in humans. For example, some investigators studying cats have reported a decrease of the late surface negative wave during SWS (Huttenlocher, 1960; Teas and Kiang, 1964) while Herz (1965) and Herz et al. (1967) reported an increase in amplitude of such a wave; this latter result was corroborated by Wickelgren (1968).

Continuous measurements of evoked responses in long, well-controlled experiments were made by Hall and Borbély (1970) at different auditory cortical depths, in the medial geniculate body, the thalamic reticular formation, and the hippocampus (Fig. 4.5). The cortical results showed that the evoked "… potentials recorded from a depth of 1–1.5 mm, all components of the average waveform being larger during slow wave sleep than in waking and low-voltage fast sleep." Responses from the cortical surface are reported including increases of the first and second positive waves as well as of the late negative wave, during SWS. The early components of both the surface and deep responses were similar in W and in PS.

The medial geniculate evoked response in sleep was represented only by an amplitude reduction during PS. Neurones of the medial geniculate nucleus showed evoked firing shifts

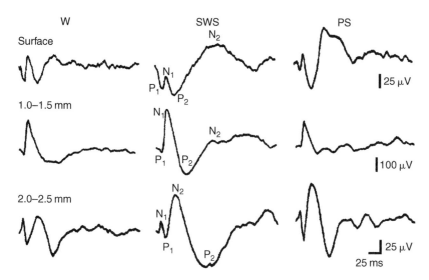

FIGURE 4.5 Click-evoked averaged cortical responses in rats, recorded from surface and from depths of 1.0–1.5 mm and 2.0–2.5 mm in the auditory cortex during W, SWS, and PS. Positive changes of potentials are indicated by downward deflections (modified from Hall and Borbély (1970)).

mainly decreasing on passing from W to SWS, while the spatial receptive field was preserved indicating that the information sent to cortical cells may carry significant content (Edeline et al., 2000; 2001).

The hippocampal and the thalamic reticular formation recordings showed increased late positive waves during SWS that switched to a smaller amplitude during both W and PS.

Thus, according to Hall and Borbély (1970), the late negative components are larger in sleep than during W in the rat, which agrees with the findings of the Herz group (1965; 1967) and also with data from Wickelgren (1968) in cats.

It has been shown (Baust et al., 1964) that middle-ear muscles contraction during PS, often accompanying rapid eye movements as phasic activity, reduced the auditory input. Although many authors did not take this possibility into account, all recording in our guinea pigs experiments were carried out without middle-ear ossicles. Besides, it was reported that when recording cats with tenotomized middle-ear muscles, the middle-ear mechanisms are not responsible for reduced medial geniculate evoked responses during PS (Berlucchi et al., 1967; Wickelgren, 1968). The hippo-campal evoked potentials reported in rats (Hall and Borbély, 1970) were quite similar to those evoked in cats; in both species the late positive wave increased markedly during SWS.

Human magnetoencephalographic evoked activity recordings in sleep

With a high temporal resolution on the order of milliseconds, the EEG and the magnetoencephalographic (MEG) are techniques with the capability to show changes of the early cortical data processing. The magnetic fields recorded from the human brain on the scalp are very small and special devices are needed to detect such signals. Furthermore, MEG has the great advantage of detecting within the millimeter range the source localization of a response, i.e., it is possible to show the dipole position over the cortex in a very precise way. This is one important finding when analyzing the responses to sound during W and sleep (Kakigi et al., 2003).

The four components of the human auditory evoked magnetic field are exhibited in Figure 4.6. The M150 and M200 waves show a clear increment in amplitude on passing from W to sleep stages I and II.

Regarding the dipole location, differences appear during sleep stages I and II; the M50 dipole wave exhibits a more anterior and lateral position on the auditory cortex. The M100 wave dipole also presents a similar change in location (Kakigi et al., 2003). The dipole cortical position shifts constitute an important change in relation to the auditory processing as well as to the sleep organization in particular, as I shall discuss later (see Chapter 5).

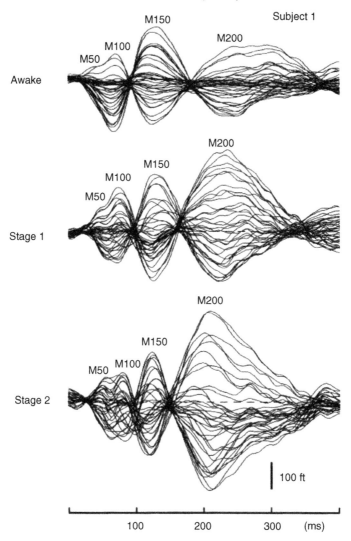

Waveforms of AEFs (250 Hz)

FIGURE 4.6   Human auditory evoked MEG response during W and sleep stages I and II. A pure tone of 250 Hz was delivered to the right ear and the magnetometer was placed on the left hemisphere (position C3). Thirty-seven superimposed waveforms at each stage are shown. Four main components (M50, M100, M150, and M200) were identified in each stage. Although M50 and M100 showed no definite change during the two stages of sleep, M150 and M200 were significantly enhanced in both sleep stages. However, M50, M100, and M200 were significantly prolonged to 1000 or 4000 Hz (modified from Kakigi et al. (2003)).

Human auditory brain areas imaging

Sensory processing in sleep have shown a new look after the introduction of the imaging studies, particularly the functional magnetic resonance technique (fMRI), although not all the approaches are univocal in their results (Braun et al., 1997; Portas et al., 2000; Born et al., 2002; Czisch et al., 2002; Tanaka et al., 2003). Portas et al. (2000) fMRI analysis exhibited the auditory stimuli in SWS producing a significant bilateral activation in the auditory cortices, thalamus, and caudate in comparison to W (Fig. 4.7). On the other hand, Czisch et al. (2002; 2004) also using fMRI, reported a reduced activation (not a lack of) in the auditory cortex during sleep stages I, II and SWS. Some of the research groups cited conclude that the decreased response may protect the sleeping brain from the arousing effects of external stimulation during sleep. These results are not congruent at all with the notion that ~50% of auditory cortex neurones in guinea pigs continue firing as during W on passing to SWS sleep, as reported by Peña et al. (1999).

In 1830, the German physiologist Burdach wrote (cited by Portas, 2005): "... an indifferent word does not arouse the sleeper, but if called by name he awakens ... the mother awakens to the faintest sound from her child ... the miller wakes when the mill stops ... hence the psyche differentiates sensations during sleep." The first researchers to demonstrate, in a controlled experiment, that personal names evoke more K-complexes in the sleeper than any other name or sounds of same intensity were Oswald et al. (1960). Moreover, the analysis of evoked potential showed a differential cognitive response to the presentation of the subject's own name – significant sounds – during sleep stage II and PS (Berland and Pratt, 1995; Perrin et al., 1999).

The fMRI analysis of data associated with EEG activity exhibited a specific response to the subject's own name during SWS sleep (Fig. 4.7). In addition, presentation of the subject's own name during sleep was associated with selective activation of

Tone-burst stimulation during sleep

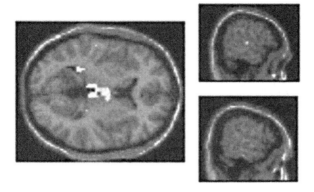

Name stimulation during sleep

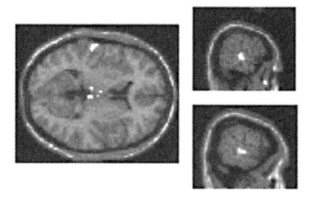

FIGURE 4.7 Functional magnetic resonance imaging. Brain areas activation both during auditory stimulation with a non-significant sound, tone-bursts, and with a significant one, the person's name, during sleep and in comparison to W (modified from Portas (2005)). See Plate 4.7 for the colour version of this figure.

the left amygdala and left prefrontal cortex (Portas et al., 2000; Portas, 2005).

In humans, the cortex activity changes are not homogeneous. When compared to W, the less active areas are located in associative cortices of the frontal and parietal cortices while the primary cortex is the least or not deactivated (Braun et al., 1997; Maquet et al., 1997; Andersson et al., 1998; Kajimura et al., 1999). The sleeping brain is a totally different brain that involves activated–deactivated areas to complete its many physiological aspects (Velluti, 2005). Sleep is a state still responsive to external stimuli (Velluti and Pedemonte, 2002; Edeline, 2003; Pedemonte and Velluti, 2005a; Velluti, 2005). Event-related potentials and also the evoked potentials – far and local-field potentials – recorded in humans and animal have demonstrated that environmental as well as internal, body, are processed in sleep.

The many results obtainable suggest that the processing of external stimuli can go beyond the primary cortices during stages I, II and SWS. The neurophysiological actions by which salient stimuli can recruit associative cerebral areas during sleep remain unclear (Maquet et al., 2005).

In comparison with W, differential patterns of regional cerebral blood flow activity are observed during stages I, II, SWS, and PS in humans. Maquet et al. (2005) suggested, "… the neural populations recently challenged by a new experience are reactivated and increase their functional connectivity during the post-training sleep episodes, suggesting the off-line processing of recent memory traces in sleep."

Auditory unit activity
in sleep

The processing of sensory information is definitely present during sleep, although profound modifications occur. All sensory systems, visual, auditory, vestibular, somesthetic, and olfactory demonstrate some influence on sleep and, at the same time, sensory systems undergo changes that depend on the sleep or waking state of the brain (Velluti, 1997). Thus, not only different sensory modalities encoded by their specific receptors and pathways may alter the sleep and waking physiology, but also the sleeping brain imposes "rules" on the incoming information. It is suggested that the neural networks responsible for sleep and waking control are actively modulated by the sensory input in order to enter and maintain sleep and wakefulness (W). Furthermore both sensory stimulation and sensory deprivation may induce changes in sleep/waking neural networks. This leads to the conclusion that the central nervous system (CNS) and sensory input have reciprocal interactions, on which normal sleep/waking cycling and behavior depend.

The present appraisal will contribute to examine the results related to how auditory sensory information is worked out by the waking and sleeping brain from a single unit viewpoint.

The auditory neurones firing rate, their temporal discharge distribution – pattern – and the relationship to hippocampus theta rhythm will be considered as components of the auditory information processing (Velluti and Pedemonte, 2002; Pedemonte and Velluti, 2005a,b; Velluti, 2005). Furthermore, considering the auditory sensory system activity as an important piece of the sleep neurophysiology enigma, the influence of sensory incoming information is postulated as an active contributor intended for sleep processes development.

The activity of central auditory neurones in sleep and wakefulness

In hearing, as well as in other sensory modalities, there are different types of mechanisms to control the sensory input: (i) Pre-receptorial actions that regulate the stimulus energy through voluntary or reflex motor activities. Head and animal's pinna movements and contractions of the stapedius and tensor tympani muscles are examples of these regulating actions. (ii) CNS actions through the efferent system. The central influence begins to be exerted at the receptor itself as demonstrated by the changes imposed on the auditory nerve compound action potential and the cochlear microphonic recordings while guinea pig is passing to sleep (see Chapter 1).

The technical approach to reach single unit recordings in the sleep–waking physiological cycle is exhibited in Figure 5.1.

Sleep changes the neuronal activity in brainstem auditory nuclei and primary cortical *loci*. Figure 5.2 presents percentages of firing changes of auditory neurones, recorded from partially restrained guinea pigs with glass micropipettes, during W and sleep stages. The extracellular unitary response in each nucleus and in the primary cortex was grouped according to the firing rate changes upon passing to sleep. Neurones could (i) decrease, (ii) increase, or (iii) exhibit no significant shifts in their firing in

Guinea pig partially restrained during W and sleep stages

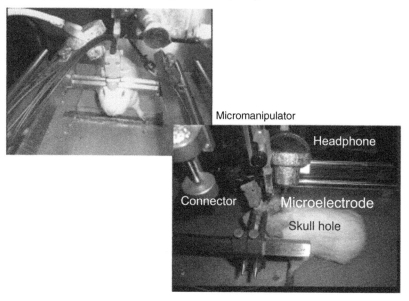

FIGURE 5.1 Guinea pigs recorded partially restricted with two metal bars attached to the skull. After recovery from surgery, these bars permit the animal head to be repositioned allowing the stereotaxic introduction of the glass micropipettes. Moreover, the body is placed in a hammock thus avoiding the head movements while allowing legs movements.

comparison to W. However, there were never completely silent neurones detected when entering into sleep, slow wave sleep (SWS) or paradoxical sleep (PS) (Peña et al., 1992; Pedemonte et al., 1994; Morales-Cobas et al., 1995; Velluti, 1997; 2005; Peña et al., 1999; Velluti et al., 2000; Pedemonte et al., 2001; Velluti and Pedemonte, 2002; Pedemonte et al., 2004; Pedemonte and Velluti, 2005a). Thus, it is concluded that the afferent input to the primary auditory cortex is not interrupted during any sleep phase, i.e., there is no auditory cortical functional deafferentation.

Approximately one half of the auditory cortical neurones responding to a characteristic frequency tone-burst maintained during sleep a firing rate equal to that observed in a previous or

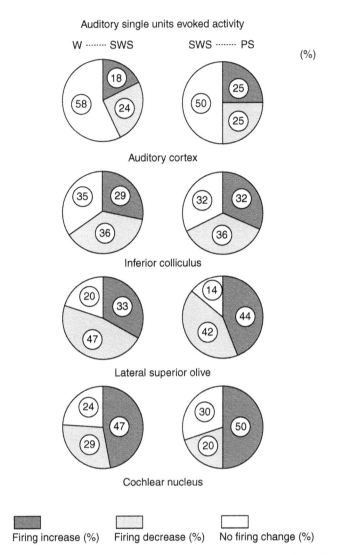

Auditory single units evoked activity

W ······ SWS   SWS ······ PS   (%)

Auditory cortex

Inferior colliculus

Lateral superior olive

Cochlear nucleus

■ Firing increase (%)   □ Firing decrease (%)   □ No firing change (%)

FIGURE 5.2   Guinea pig percentages unitary evoked activity along the auditory pathway in the sleep–waking cycle. Pie charts show percentages (%) of neuronal firing shifts on passing from W to SWS and from SWS to PS. The subcortical nuclei, inferior colliculus, and the lateral superior olive exhibited a higher percentage of increasing–decreasing firing neurones. The cochlear nucleus presented an increased activity perhaps according to the higher blood flow reported by Reivich et al. (1968). Around 50% of the cortical neurones responded as during W. No silent auditory neurone was detected on passing to sleep or during sleep in any pathway level (data from Peña et al. (1992; 1999); Pedemonte et al. (1994); Morales-Cobas et al. (1995)).

subsequent W. The other ~50% was divided into neurones that: (a) increased or (b) decreased their firing rate. This second set of neurones – (a) and (b) – may be part of neuronal networks that, in some unidentified way, could actively participate in sleep processes. Furthermore, I am supporting the notion that the incoming sensory information is functionally active in sleep processes, not only as a passive participant (Velluti, 2005), being a new approach for potential research.

The effects of sleep and W on auditory evoked activity of cats' single brainstem neurones, i.e., units at the mesencephalic reticular formation, was reported by Huttenlocher (1960). The activity of the non-lemniscal neuronal auditory pathway was shown to vary between sleep and W. The evoked activity of the units (about 50%) increased during quiet W and diminished during SWS. Approximately 30% of the recorded neurones showed evoked responses during SWS with the same activity, or even greater, than that during W. The most constant effect on evoked neuronal firing, as well as on spontaneous activity, was observed during PS in cats. During this sleep phase the responses to clicks diminished in the cells studied ($n = 16$) while in some neurones ($n = 5$) the activity evoked by the click was completely absent. The firing returned upon arousal (Huttenlocher, 1960).

Four examples of changes in the unitary response on passing from W to SWS and PS are shown in Figure 5.3.

A cochlear nucleus example neurone that exhibits a "primary-like" post-stimulus time histogram increased its firing during SWS and PS. The control W periods, pre- and post-sleep epochs, showed a remarkable W pattern and firing similarities.

In the anteroventral cochlear nucleus of the guinea pig extra-cellularly recorded responses revealed neurones having clear-cut relationships with the sleep–waking cycle (Velluti et al., 1990; Peña et al., 1992). In agreement with previous results yielding increments in the averaged auditory nerve compound action potential amplitudes during SWS (Velluti et al., 1989), 72% of the cochlear nucleus units spontaneously discharging and 47% of the

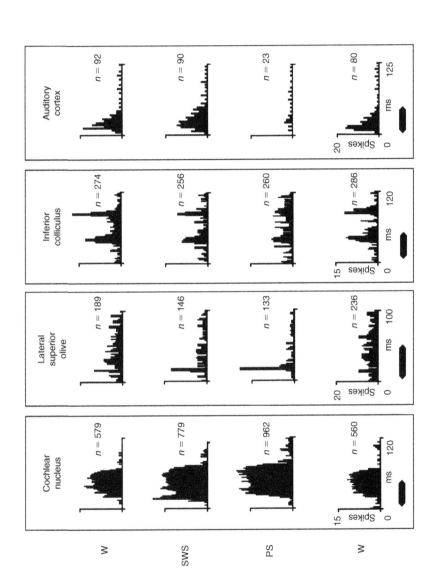

units responding to sound increased their firing in SWS. Some units (27%) did not show any firing change as compared with W. During PS, 20% of sound responding and 66% of spontaneously firing neurones followed the auditory nerve amplitude trend while, on the other hand, a great proportion of units failed to follow the auditory nerve pattern. There were also observed differences in the probability of discharge over time – pattern – between the sleep phases and W (Fig. 5.4).

It has been postulated that the auditory efferent system modulates the auditory input at the level of the cochlear nucleus during sleep and W. The probability of firing and the changes in the pattern of discharge are thus dependent on both the auditory input to the cochlear nucleus and the brain functional state, e.g., asleep or awake (Fig. 5.4; Peña et al., 1992). Direct descending connections from cortical *loci* were also reported (see Chapter 1).

The lateral superior olive example presents a different situation: the example unit progressively decreases its discharge rate when passing to sleep phases while the pattern of discharge changes neatly (Fig. 5.3). The neurone that initially, during W, was an "on-sustained" unit is transformed into an "onset" one during PS, i.e., discharging only at the beginning of the stimulus. Most neurones from the guinea pig lateral superior

FIGURE 5.3   Four representative auditory neurones recorded at different auditory loci (cochlear nucleus, lateral superior olive, inferior colliculus, and auditory cortex) during the sleep–waking cycle are shown. Post-stimulus time histograms exhibit changes in the pattern and/or in the frequency of discharge on passing from W to SWS and PS. In these examples the cochlear nucleus recorded neurone increases the firing rate during sleep maintaining the same pattern of discharge; the lateral superior olive shows both a change in pattern and a decrease in firing during sleep. The inferior colliculus neurone exhibits a changed pattern but not significant variation in firing rate. The auditory cortex neurone significantly decreases its firing only during PS, recovering it in the following W epoch. Stimuli: tone-burst (50 ms, 5 ms rise-decay time, at the unit characteristic frequency) (data from Peña et al. (1992; 1999); Pedemonte et al. (1994); Morales-Cobas et al. (1995)).

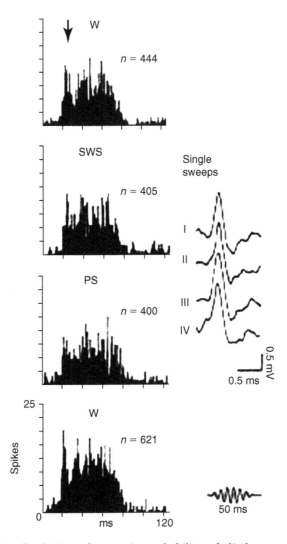

FIGURE 5.4 Qualitative changes in probability of discharge over time –
pattern – of a cochlear ventral nucleus auditory neurone during W and sleep
stages. The upper W epoch shows a dip of decreased firing probability (arrow)
in the post-stimulus time histogram. During the successive episodes of SWS
and PS neither the firing nor the pattern change significantly although with a
different probability, during W the firing dip disappears. The last W epoch –
after sleep – shows both a discharge increase ($n = 621$) and a pattern shift, exhibiting a "chopper" type of response, a quite different pattern. On the right side,
individual sweeps are shown (I–IV) demonstrating the same unit was always
recorded. Unit characteristic frequency: 0.85 kHz; 48 dB SPL (modified from
Peña et al. (1992)).

olive showed firing rate modulation on passing from W to SWS; 80% of the recorded cells changed their firing during binaural stimulation while 85% did so during ipsilateral sound stimulation. In addition, shifts in the discharge pattern were observed in 15% of the cells recorded on passing from W to sleep. The most striking change was observed associated with the decreasing firing units: a very low discharge number, after a peak of responses in PS, during the last 40 ms is shown in the post-stimulus time histogram (Fig. 5.3).

The waking cues for binaural directional detection – in this particular experimental paradigm – disappeared during SWS; one possible interpretation of this result is that the binaural function of some (11.5%) lateral superior olive cells is impaired during SWS (Fig. 5.5).

Auditory efferent pathways are impinging on the lateral superior olive neurones, thus, their activity is dependent on ascending input as well as on cortical descending influences, the latter being behaviorally dependent, sleep and waking in our case (Pedemonte et al., 1994).

The example neurone of the inferior colliculus changes the pattern, i.e., the temporal distribution of the neuronal discharge, in spite of a non-significant firing number shift, on passing from W to sleep phases (Fig. 5.3). Most neurones (63%) exhibit evoked firing rate increases or decreases on passing from W to sleep. The majority of the inferior colliculus units shift their evoked firing during PS while only 11% did not show such changes. Moreover, inferior colliculus cells spontaneous discharge is observed to increase in most units during PS (Morales-Cobas et al., 1995). Auditory information processing is present during sleep stages although with differences regarding W (Fig. 5.6).

The inferior colliculus auditory neurones send descending connections to regions such as the dorsal pontine nuclei, known to mediate sleep signs and/or processes, making this *locus* suitable for sleep–auditory system interactions. In addition, both the cochlear and trapezoid nuclei are sources of abundant

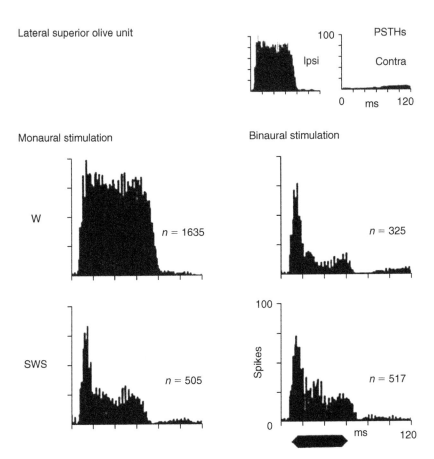

FIGURE 5.5 Tone evoked responses from a lateral superior olive neurone during waking (W) and SWS, with monaural and binaural sound stimulation. Upper traces show the unit response to ipsi- and contralateral auditory stimuli (45 dB SPL; characteristic frequency 1.0 kHz) exhibiting neat differences. During W, monaural and binaural sound stimulation showed great changes including pattern and firing probability ($n = 1635$ versus $n = 325$). On passing to SWS the neurone exhibited lower firing ($n = 505$) with monaural stimulation, the same pattern and very similar discharge number with binaural stimulation; this neurone and its neuronal network cannot recognize the difference between monaural and binaural sound stimulation in this sleep stage, SWS (modified from Pedemonte et al. (1994)).

Inferior colliculus central nucleus

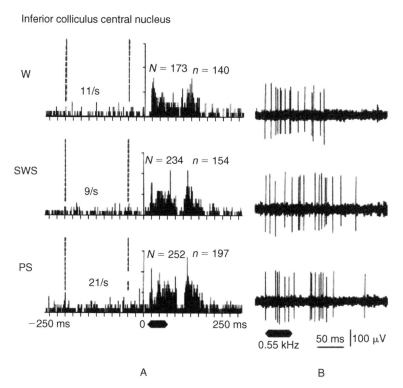

FIGURE 5.6   Evolution of an inferior colliculus auditory unit firing through-out the sleep–waking cycle. In (A) the peristimulus time histograms (PSTH) exhibited an increased evoked firing (N) on passing from W to SWS and PS. The second half of the PSTH (n) showed increasing firing on passing from W to SWS, while when changing from SWS to PS, another discharge decrease was present. In (B), a unit raw recording is shown exhibiting the split discharge pattern observed in the PSTHs. On the other hand, the spontaneous firing rate (PSTH left side) presented a slight firing decrease on passing from W to SWS while a highly significant ($p < 0.001$) increase was obtained during PS (modified from Morales-Cobas et al. (1995)).

projections to the rostral part of the nucleus *reticularis pontis caudalis* (Garzón, 1996), which in turn is strongly connected with the PS signs induction zone located in the ventral part of the nucleus *reticularis pontis oralis* (Reinoso-Suárez et al., 1994).

Neurones of the medial geniculate nucleus, the auditory thalamus, showed evoked firing shifts mainly decreasing on passing from W to SWS, while the spatial receptive field was preserved indicating that the information sent to cortical cells may carry significant content (Edeline et al., 2000; Edeline, 2003; Hennevin et al., 2007).

The auditory areas of the cerebral cortex receive complex ascending input, originating from both ears and, in turn, send projections to thalamic and midbrain targets (Saldaña et al., 1996). Thus, the auditory cortex may control the whole auditory system and, at the same time, be dependent on the general state of the brain, asleep or awake. In the guinea pig, auditory cortical units recorded at their characteristic frequency varied their firing evoked activity on passing from W to sleep phases. The firing shifts were observed during SWS, exhibiting decreased–increased firing while a percentage did not show firing rate changes (Fig. 5.2). Extracellular recordings of both spontaneous and sound evoked responses from the primary auditory cortex (A1) have been shown to be highly dependent on the state of the brain, e.g., W, SWS, or PS (Fig. 5.3). Thus, the sleep and W brain states modify the auditory cortical processing of simple sound stimuli (Peña et al., 1999).

An important result in the Peña et al. (1999) study is the divergent change obtained between spontaneous and evoked unitary activity. The trend was not always the same; the evoked activity could increase while the spontaneous activity decreased or *vice versa*. A hypothesis that emerges from these results is that there are at least two separate controls exerted on the same neurone: one correlated with the incoming signals and a second control perhaps from a non-auditory CNS locus and from stored auditory information *loci*. As reported in the primate visual cortex, the secondary visual areas increased their activity during PS when studied with positron emission tomography (Braun et al., 1997; 1998). Two questions readily arise: first, was the neuronal activity in those visual areas related to the visual imagery normally occurring during PS? And second, was the increased spontaneous activity in our auditory cortical results, particularly during PS, also related to auditory hypnic "images"?

It is remarkable that in our results a number of the neurones in the primary auditory cortex exhibited no significant quantitative changes in their evoked or spontaneous firing rates that could be correlated to behavioral, brain states shifts. This is inconsistent with reports in the literature but consistent with the hypothesis that, to a varying extent, the responsiveness of the auditory system is preserved during sleep. However, an analysis in which the temporal development of the neuronal discharge is studied over the sleep/wakefulness cycle remains to be done. Nevertheless, the present study demonstrates that significant changes can be correlated to behavior during either spontaneous or evoked activity in 42.2–58.3% of the sampled neurones. Such auditory cortical neurones – and the neural networks to which they belong – exhibit a different behavioral state-related functioning, probably reflecting a general shift in the brain sensory processing during sleep. About half of the auditory cells were as active as during W on passing to SWS or PS. This cortical region is not disconnected from the outer world, on the contrary, it is perhaps monitoring sounds from the environment.

Neuronal discharge pattern shifts

The firing pattern change may support a different possibility of sound analysis as well as suggest a different mode of relation to other cell assembly/network which I am herein postulating as related to active sleep processes. The same neurone may exhibit a pattern during SWS and a different one during PS, to recover the initial firing distribution at the following W epoch. Moreover, diverse patterns could be observed throughout the sleep–waking cycle (Velluti and Pedemonte, 2002) perhaps in relation with the association to another, different, neuronal network. Besides, to

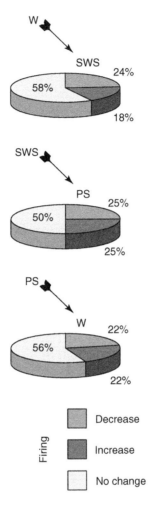

FIGURE 5.7 The pies chart shows the firing evolution of an auditory cortical neurone when passing from W to sleep and again into W.

further demonstrate how the auditory neuronal firing changes at the cortical level on passing from W to SWS and PS, and back from PS to W, Figure 5.7 shows this dynamic condition during a full sleep cycle.

Brain's maths

Neural models predict that mathematical operations take place in neurons or systems of neurons. How realistic such operations are is a matter of debate. The auditory system offers a good substrate to approach this question because it is possible to test how neurons respond to the manipulation of acoustic signals. In order to determine what and where acoustic objects are, our brain must compare and combine the noisy information that arrives to each ear. Comparison, combination, and noise reduction can be represented by equations. A common mathematical tool to extract the critical signal out of a highly variable set of stimulus is averaging. Similarly, cross-correlation and multiplication are mathematical correlates of comparison and combination selectivity, respectively. In the avian auditory pathway, we can find neurons that appear to perform each of these types of processing.

The neurons of the nucleus laminaris, the avian equivalent of the medial nucleus of the superior olive, perform a running cross-correlation of signals coming from each ear. This computation allows neurons to indicate the degree of similarity between the right- and left-side inputs necessary to identify what sound is coming from a unique source. The space-specific neurons that constitute the map of auditory space in the avian midbrain are selective to combinations of spatial cues. This combination selectivity, which must be tolerant to a broad range of sound intensities, is implemented by an effective multiplication of inputs from independent processing pathways. Multiplication not only makes the response to particular combinations larger but also lends veto power to one spatial cue over the other. In the central nucleus of the inferior colliculus, neurons achieve a tuning to the horizontal coordinate of sound direction that would require at least five repetitions of the same stimulus in lower-order nuclei. This increase in reliability is due to an average-like process that takes place in a single step. Thus, the hierarchically organized auditory system seems to implement the computations that models predict. To this point, however, we cannot go beyond saying that we have found conditions under which these operations describe the neurons' behavior. Nor can we assume that these response properties are identical in behaving animals, as these studies have been performed under anesthesia. Yet, it makes us wonder about how close to our nature mathematical operations are.

José L. Peña
Yeshiva University
New York, USA

This is indicative of a neatly reversible physiological process, being also another experimental demonstration that the afferent input to the auditory cortex is dynamic, although predictable, and that the auditory cortex is not functionally deafferented during sleep. In the primary auditory cortex (AI) example, a continuously changing unitary firing is shown on passing from W to SWS, from SWS to PS, and from PS to W. The firing recovery in the subsequent W after a PS epoch once a sleep cycle is completed, exhibits the relative constancy of the phenomenon (Fig. 5.7).

Bursting units firing at the auditory cortical level are not, or in very few cases, present during W or sleep. The proportion of such bursting neurones in anesthetized states is much larger (Fig. 5.8), i.e., it gives the impression of an artifact phenomenon rather than a physiological one. In my experimental data it was never recorded a bursting cell at the cortical level during W or sleep in guinea pigs.

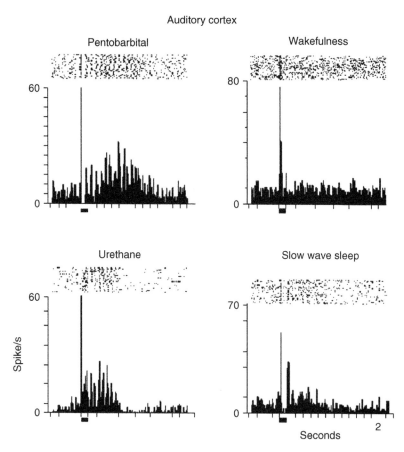

FIGURE 5.8   Tone evoked oscillations are present in the post-stimulus time histograms (PSTH) of anesthetized animals but not in non-anesthetized, e.g., awake or asleep animals. Although the PSTH and raster exhibit oscillations under pentobarbital or urethane anesthesia, there is no tendency for such oscillating pattern in waking or SWS (modified from Edeline (2005)).

Relationships between the theta rhythm of the hippocampus and auditory neurones

The hippocampus produces a high-amplitude transient theta rhythm when, e.g., a cat is looking itself in a mirror; this

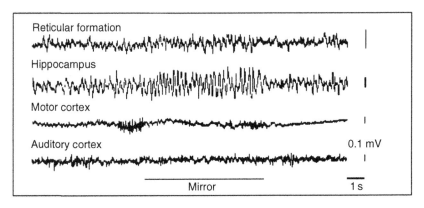

FIGURE 5.9   Recording from a wake cat showing enhancement of theta rhythm in the hippocampus when the animal sees itself at a mirror (bar) (modified from Grastyan et al. (1959)).

observation by Grastyán et al. (1959) connected this rhythm to a sensory input, the vision, and to higher processes as, perhaps, conspecific recognition (Fig. 5.9). At the same time, recordings carried out in the primary auditory cortex showed evoked neuronal firing shifts elicited by electrical stimulation of the hippocampus, indicating that these brain regions are interconnected exhibiting a functional relationship. These results support the notion that an auditory-hippocampus shared functional interaction, although unknown in detail, may be present (Cazard and Buser, 1963; Redding, 1967; Parmeggiani and Rapisarda, 1969; Parmeggiani et al., 1982; Buszaki, 1996).

The ultradian hippocampal theta rhythm, within the wakefulness–sleep circadian cycle, may modulate the sensory neuronal activity. It has been one of the conspicuous time givers postulated as an internal *zeitgeber*, a temporal organizer for auditory sensory processing (Pedemonte et al., 1996b; Velluti et al., 2000; Velluti and Pedemonte, 2002; Pedemonte and Velluti, 2005b). The theta waves, 4–10 cycles per second (cps), may affect spatially distant neurones by inducing fluctuations of the cellular excitability due to membrane potential oscillations (García-Austt, 1984; Kocsis and Vertes, 1992). Although more prominent in active W and PS, the theta rhythm is always present in the brain, e.g., the hippocampus theta frequencies can also be observed during SWS when analyzed in the frequency domain, the Fourier transform (Komisariuk, 1970; Gaztelu et al., 1994; Pedemonte et al., 1996b).

This particular hippocampus rhythm has been related to several brain processes; it was found involved in motor activities during both W and PS (Buño and Velluti, 1977; García-Austt, 1984; Lerma and García-Austt, 1985), as well as in the sensory processing related to a motor context (Grastyán et al., 1959; Kramis et al., 1975). It was also found implicated in spatio-temporal learning (Winson, 1978; O'Keefe and Recce, 1993), associating distant, discontiguous events (Wallenstein et al., 1998) and learning of temporal sequences (Mehta et al., 2002). A role of the theta rhythm in learning and memory has been proposed from different viewpoints (Adey et al., 1960; Burgess and Gruzelier, 1997; Doppelmayr et al., 1998; Klimesch, 1999; Kahana et al., 1999; Vinogradova, 2001), and it has also been linked to the modulation of autonomic processes such as the heart rate in the guinea pig and human (Pedemonte et al., 1999; 2003).

The auditory neurones exhibit phase-locking, i.e., temporal correlation with theta rhythm at different levels of the auditory pathway, from the brainstem – the cochlear nucleus, the superior olive, the inferior colliculus – to the primary cortical region (Pedemonte et al., 1996b; Velluti et al., 2000; Velluti and Pedemonte, 2002; Pedemonte and Velluti, 2005b). The processing in other sensory modalities has been associated with the theta rhythm, e.g., touch (Nuñez et al., 1991), pain (Vertes and Kocsis, 1997), vision (Gambini et al., 2002), and olfaction (Margrie and Schaefer, 2003; Affani and Cervino, 2005). Thus, it is my tenet that the theta rhythm of the hippocampus contributes, in the time domain, to the complex dilemma of the central auditory information processing in sleep and waking.

Most lateral superior olive auditory neurones showed a spontaneous firing rate modulation on passing from W to SWS

Rhythms and auditory neuronal activity

One of the most exciting experiences of *in vivo* electrophysiology is to watch a neurone discharging online. The neuronal decision to fire or not an action potential can integrate information carried by thousands of synaptic inputs at a given time. Central auditory neurones can discharge tens of spikes per second, being not all of them synchronized with sound stimuli.

Because of the auditory units firing differences observed between W and sleep, the question is what information does the neurones that shift firing in sleep carry? Perhaps, those units are related to some sleep processes.

Research in audition deals mainly with the study of neuronal responses to pure tones, white noise, and artificially synthesized acoustic signals. However, the real auditory world consists of complex combinations of frequencies, whose power is distributed along time in specific ways, and this world is interpreted by the brain according to its current state, waking or asleep. Getting to know how the brain processes that information is perhaps the most important challenge for neuroscientists.

The brain presents many spontaneous rhythms, which can be recorded as field potentials. These rhythms are the result of internal processes that change with states such as sleep, W, attention. Thus, I am speculating, brain rhythms must be involved in information processing under different conditions. Auditory information needs a temporal frame for events to be analyzed in particular sequences. Words and music have sense if they are perceived along time. Among the variety of rhythms that the brain offers, the theta rhythm (4–12 cps) could be a good candidate because: (1) it is present in any behavioral state; (2) it is generated, amplified, and distributed by a structure involved in learning and memory processes, i.e., the hippocampus; and (3) it has been related to auditory and visual sensory systems, motor and autonomic processes during sleep and W. Furthermore, studying mice, it has been established that theta frequency in PS is controlled by a single autosomal recessive gene (Tafti 2003), suggesting that kind of genes may be found for rhythms variant during human sleep.

Over more than a decade of research, our group found a temporal correlation between the hippocampal theta rhythm and the discharge of auditory neurones, postulating that the theta rhythm is a timer in auditory processing, which lends experimental support to the hypothesis of this rhythm as a time giver (Pedemonte et al., 1996a; Pedemonte and Velluti, 2005b).

What happens with the human being? Many human functions are rhythmic, e.g., respiration, heart beating, automatic movements (walking, running, swimming, etc.). It is not only easier to learn rhythmic tasks but humans are attracted to sounds that have a particular *tempo*. Have you ever observed the people's behavior in a music concert? You can see them performing enthusiastic rhythmic movements, often synchronized with frequencies within the delta and theta range. When music becomes slower or faster, people abandon the synchronized movements.

In conclusion, brain rhythms – that change during sleep and W – could offer different frequency ranges for the temporal, and perhaps also spatial, organization of the brain auditory processing.

Marisa Pedemonte
CLAEH Instituto Universitario
Punta del Este, Uruguay

and PS (Fig. 5.10; Velluti et al., 2000). The interactions between the auditory units and the hippocampal theta, demonstrated by the cross-correlations, exhibited phase-locking during W and PS, at the same hippocampal theta wave phase. In this example,

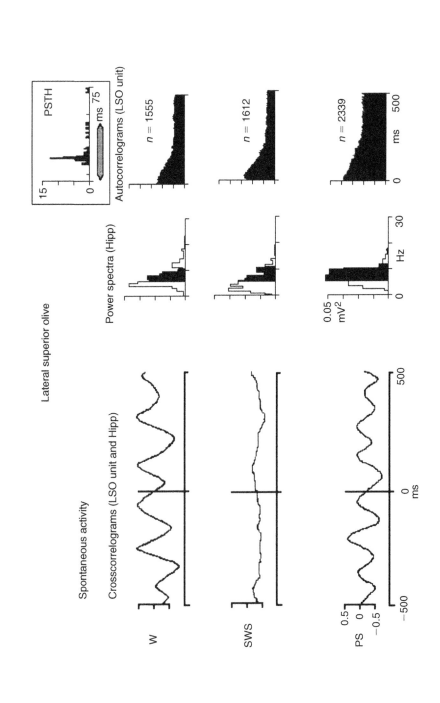

Lateral superior olive

Spontaneous activity

Crosscorrelograms (LSO unit and Hipp)

Power spectra (Hipp)

Autocorrelograms (LSO unit)

PSTH

W

SWS

PS

$n = 1555$

$n = 1612$

$n = 2339$

ms 75

15

0

0.5
0
−0.5

0.05
mV²

−500   0   500
ms

0   Hz   30

0   ms   500

during SWS, the phase-locking failed to appear in spite of a hippocampal theta occurrence shown in the corresponding power spectrum (Fig. 5.10). In other unit recordings the phase-locking was also present during SWS.

Inferior colliculus neurones showed a higher synchrony with hippocampal theta when sound stimulated at the unit's characteristic frequency during waking, although the spontaneous activity also exhibited phase-locking (Fig. 5.11). Shifts in the angle of phase-locking to the theta rhythm were also observed during PS. Moreover, when a continuous pure tone (at the neurone characteristic frequency) was delivered, the single unit response became rhythmic at the theta frequency (Pedemonte et al., 1996b). During PS all inferior colliculus auditory neurones recorded exhibited hippocampal theta correlation: 40% were rhythmic and phase-locked to the theta frequency and 60% were non-rhythmic maintaining the theta phase-locking (Fig. 5.12).

The hippocampal formation is a high-level cortex specifically involved in memory encoding. Moreover, it has been suggested that the hippocampal circuitry organization serves as a matrix for rapid auto-associative memory (McNaughton and Morris, 1987). Sounds such as a word, a musical phrase, etc., develop in temporal sequences, so a special rapid memory and temporal organization for receiving the signals may be necessary. Furthermore, the hippocampus "... appears as a sort of holding system that is necessary for the temporary storage of information regarding

FIGURE 5.10   Functional relationship between a lateral superior olive neuronal spontaneous discharge and hippocampal (Hipp) theta rhythm during behavioral states. The auditory unit firing is phase-locked with the Hipp theta rhythm during W and PS. In the example shown, the correlation is lost during SWS. On the other hand, the firing rate increases during SWS and particularly during PS (autocorrelogram). The power spectra exhibit a strong theta frequency band power in each state. Inset: post-stimulus time histogram (PSTH) of the unit; binaural sound stimulation (tone-burst at characteristic frequency: 2.0 kHz; 48 dB SPL) (modified from Velluti et al. (2000)).

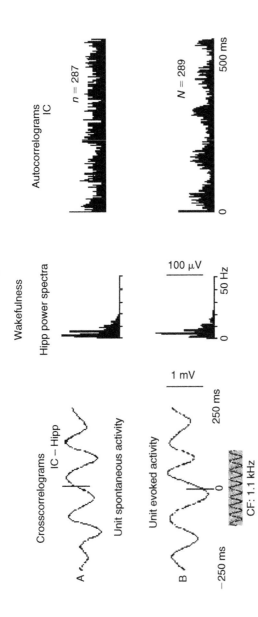

FIGURE 5.11 Temporal correlation between hippocampal (Hipp) theta rhythm and inferior colliculus (IC) auditory single unit spontaneous and evoked activity. Both, spontaneous and evoked firing showed cross-correlations (CC) – phase-locking – with Hipp (bipolarly recorded) theta rhythm. The phase-locking showed a small timing difference between the CC with spontaneous and evoked firing. The autocorrelation histograms (AC) exhibited that the non-rhythmic discharge pattern during spontaneous activity (upper trace) becomes rhythmic at theta frequency during the evoked activity (lower trace), thus stressing the relevant influence of the auditory input on the theta waves. The correlation of the spikes with the theta rhythm was studied with spike-triggered averaging of the Hipp field activity. Continuous pure tone sound stimulation at the unit's characteristic frequency (CF): 1.1 kHz; 10 dB above threshold (modified from Pedemonte et al. (1996a)).

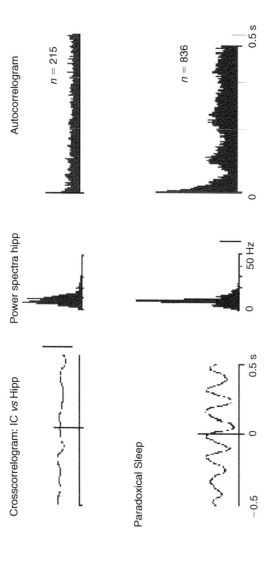

FIGURE 5.12 Cross-correlation of CA1 hippocampus theta rhythm and inferior colliculus auditory neurone sponta-neous activity during W and PS. This neurone behaved in a non-rhythmic manner without phase-locking to the theta rhythm during W. The autocorrelogram also shows this unit as non-rhythmical. In the PS stage, it spontaneously became both rhythmic and phase-locked to the hippocampal theta rhythm and the unit firing increased. The theta rhythm fre-quency was enhanced around the 6–7 c/s band. Cal.: Crosscorrelogram, 1 mV; power spectra bar, 50 μV (modified from Pedemonte et al. (1996a)).

the temporal order and the spatial context of events" (Rawlins, 1985; Lopes da Silva et al., 1990). Finally, the hippocampal theta rhythm may play this role as an internal clock (*zeitgeber*), adding a temporal dimension to the processing of auditory sensory information (Velluti and Pedemonte, 2002).

Furthermore, the hippocampus is involved in the neural coding of spatial positions (O'Keefe and Recce, 1993; Wallenstein et al., 1998; Best et al., 2001) and it is necessarily associated with sensory processing to establish its space location. As an experimental animal traverses space, the hippocampal place neurones firing progressively change to an earlier phase of the ongoing theta rhythm (Skaggs et al., 1996; Magee, 2003).

In agreement with the writer Jorge L. Borges (1899–1986), "Movement, that is occupying different positions ... is unconceivable without time, ..."; time, not only space, is the other variable that may be controlled by the hippocampus and it is represented by the field activity, the theta rhythm, which is postulated as a meaningful factor in the temporal processing of auditory signals. The temporal correlation phenomenon – phase-locking – was described during W as well as in both sleep phases, SWS and PS (Pedemonte et al., 1996b; Velluti et al., 2000; Pedemonte et al., 2001; Velluti and Pedemonte, 2002).

The phase-locking is not a fixed phenomenon that may change associated to known and unknown factors. For instance, when a sound appears or a different sensory stimulation is introduced in order to change the animal attention, it is possible to evoke the theta phase-locking and the increment of theta power for many seconds (Fig. 5.13). These shifts may occur most of the time in W as well as during all sleep stages in the inferior colliculus auditory recordings (Liberman et al., 2006). Furthermore, lateral geniculate visual neurones may evoke unitary discharge phase-locking when changing the light flashes frequency (Pedemonte et al., 2005). Besides, a cyclic on and off temporal correlation to the theta rhythm of about 5 seconds has been observed in previous reported data similar to our results

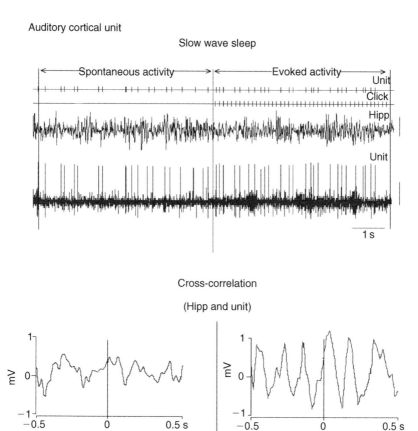

FIGURE 5.13   Relationship between hippocampal (Hipp) theta rhythm and auditory cortical unit during SWS. Upper traces: raw data showing – from top to bottom – digitized units, sound stimuli (clicks), Hipp field electrogram, and auditory cortical unitary discharges. The spontaneous activity is shown on the left of the vertical line and, on the right part, the activity during sound stimulation (8/s). The crosscorrelogram did not exhibit phase-locking with the Hipp theta rhythm during the auditory unit spontaneous discharge. However, when the sound stimulation started, the neurone began to fire in close correlation with a particular theta rhythm phase, phase-locked. Cal.: Hipp, 1 mV; unit, 50 μV (modified from Velluti et al. (2000)).

that change the attention level during ~5 seconds (Vinogradova, 2001; Velluti and Pedemonte, 2002). Vinogradova (2001) classifies the modulating influence of theta rhythm in two systems: (i) a regulatory system, linking the hippocampus to the brainstem structures, which senses the attention level, introducing primary information about changes in the environment and (ii) an informational system which holds reciprocal interactions with the neocortex.

Human intracranial recording has revealed theta oscillations in cortical places, suggesting the existence of theta generators in the brain surface. Theta waves were observed in the basal temporal lobe and frontal cortex without a functional coupling between neocortex and hippocampus during theta periods. This is indicative of multiple theta generators in human brain and evolving from tonic – in lower mammals – to phasic in PS (Kahana et al., 2001; Cantero et al., 2003).

Although a temporal correlation between the Hipp theta and the auditory cortex unitary activity does not necessarily imply a direct influence of one over the other, the conspicuous correlation of Hipp theta and several other neural processes suggests the existence of some kind of interaction (Fig. 5.14). This correlation has been observed during W and also during both sleep phases. It is consistent with the proposed hypothesis that Hipp theta rhythm supplies a temporal dimension to the processing of auditory information and may be acting as an internal clock as previously suggested (Adey et al., 1966; Pedemonte et al., 1996a; 2001; Velluti et al., 2000). The shuffle of the spike series supports the statistic validity of the results because there was no phase-locking between the cortical units and theta waves after data "shuffling." The phase-locking may depend or not on the power of theta hippocampal field potential. On changing behavioral state, a temporal relationship – phase-locking – was found during W, SWS, and PS. Besides, this correlation may shift when neurones are acoustically stimulated and, the same neurone could show different correlation for the spontaneous and evoked activities. The

Brain plasticity versus brain homeostasis

One of the most basic concepts underlying contemporary neuroscience is that the adult mammalian brain changes every time that a new motor or cognitive ability is acquired and stored. However, this basic tenet is supported on several assumptions that need to be stated clearly. First, as part of a changing universe, the brain is always changing. We, as neuroscientists, should distinguish which changes are spontaneous and which represent the substrate of learning acquisition and storage. Second, even when the brain changes either spontaneously or due to its own activity-dependent plasticity, humans are still capable of maintaining concepts, ideas, and beliefs for relatively long periods of time. Finally, the brain can also suffer significant structural changes in the absence of any noticeable behavioral or cognitive changes, as in the early asymptomatic periods of many neurodegenerative diseases or during some regenerative processes. Thus, it seems the brain permanently faces a decision between plastic versus stabilizing processes.

The term brain plasticity is not necessarily synonymous with brain functioning. Regarding plasticity, we assume the presence of some structural, albeit localized and subtle, neural changes underlying learning and memory processes. But the brain is also able to be in different states – called asleep, awake, or dreaming – involving changes like receiving sensory information that are not necessarily structural, rather functional. Thus, it is still possible that learning mechanisms are ascribed to the dynamic, emergent properties of neural ensembles. We have more neurons than proteins, and perhaps the former can carry out a good job without the need of any structural modifications of their already sophisticated connectivity. Why, then, do most neuroscientists prefer to lean on *neural plasticity* rather than on *neural functional states*? The most parsimonious answer is that we have collected a huge amount of information about the structure and connectivity of neural tissue at subcellular and molecular levels, and about the anatomical and biochemical rules and pathways maintaining these structures and circuits. In addition, definite behavior and sensory-motor properties are easily ascribed to specific neural sites. In contrast, our information about brain functioning during learning situations is too constrained by the limitations imposed by electrical recordings from small numbers of neural elements selected out of billions, or by modern mapping techniques dealing with electrical or biochemical representations of brain activity.

Although re-routing of given neural connections is proposed from time to time, none has been convincingly demonstrated in the brain of adult mammals. In fact, it has never been proven that post-embryonic peripheral or central neurones are able to re-specify their functional codes following their reinnervation of new neural or muscle targets. Plasticity is then restricted to the local micro-environment surrounding postsynaptic neurones, although molecular processes involving presynaptic terminals cannot be ruled out. These structural changes are normally described as the final steps of functional changes named long-term potentiation and long-term depression. Thus, the search for the neural engram or learning trace (either localized at the hippocampus or properly distributed in the neural tissue) has now been replaced by the search for a (mostly intracellular) micro-engram. In accordance with this generally assumed, contemporary trend, selected brain sites should be endowed with the necessary and sufficient molecular machinery to produce the pre- and/or postsynaptic changes supporting the acquisition, storage, and retrieval of motor and cognitive learning (Delgado-García and Gruart, 2002).

José M. Delgado-García
Universidad Pablo de Olavide.
Sevilla, España

influence that attention processes exert on hippocampal activity may indicate a point of interaction between those processes and the changes in the pattern of discharge of auditory neurones in sleep and W. Our results are indicative of a new approach to sensory processing analysis in relation to behavioral states and particularly with all sleep stages.

The sleep–waking dissociation between evoked and spontaneous unitary activity was reported by Huttenlocher (1960) for auditory units recorded from the mesencephalic reticular formation. Examples of a divergent control between spontaneous and evoked firing rate were reported at auditory nuclei as well as at cortical level suggesting a differential central control mode (Morales-Cobas et al., 1995; Peña et al., 1999). The phase-locking differences observed between spontaneous and evoked discharges are consistent with the existence of multiple synaptic inputs converging over the neurone. Thus, the observed differences in the Hipp theta unit phase-locking of spontaneous and evoked activity, reported by Pedemonte et al. (2001), is also suggestive of a different mode of central control.

We may assume that the same cell may be active during different behaviors although engaged with different networks. Besides, it is well known that hippocampal place units activity represent an environmental map associated to behavior and sensory input as demonstrated by O'Keefe and Recce (1993). Also, a temporal and/or spatial positioning was reported by Wallenstein et al. (1998) for hippocampal units. Both approaches suggest the necessity of a close temporal control of the input signals that could be supplied by the Hipp theta rhythm, a low-frequency *quasi*-sinusoidal rhythm.

The demonstration of a correlation between the incoming information and Hipp theta rhythm opens a new way to study the cortical processing, not only in W periods but also during sleep. The relationship established between neuronal discharges and Hipp theta rhythm may represent a possible link between attention mechanisms, W/sleep and auditory processing or, at

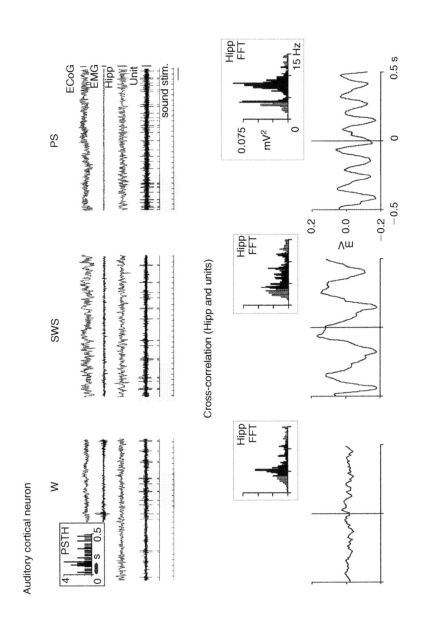

Auditory cortical neuron

W SWS PS

ECoG
EMG
Hipp
Unit
sound stim.

PSTH

Cross-correlation (Hipp and units)

Hipp FFT
Hipp FFT
Hipp FFT

0.075
mV²
0

15 Hz

0.2
0.0
−0.2
mV

−0.5 0 0.5 s

least, a temporal input organization, supporting the notion of an internal clock role in W and, most intriguing, during sleep (Pedemonte et al., 2001).

Complex sound processing at the auditory single unit level

Few experimental works have been conducted in sleeping animals even with artificial, simple stimulus, while none has employed complex sounds, e.g., the natural animal call.

Different neuronal types, responding to sound, are present in the inferior colliculus and the auditory cortex. For instance, Tammer et al. (2004) reported diverse grouping of neurones related to natural sound, vocalizations, in the squirrel monkey inferior colliculus. Type 1 neurones are activated by self-produced vocalizations as well as vocalizations of group mates and non-species-specific sounds. Type 2 neurones are activated by vocalizations of group mates and other acoustic stimulus although not by self-produced calls. Type 3 units are activated by self-produced vocalization, but not by group calls or other external sounds (Tamer et al., 2004).

FIGURE 5.14   Temporal correlation – phase-locking – between the evoked discharges of a primary auditory cortex neurone and hippocampus (Hipp) theta rhythm, during sleep–wakefulness cycle. Top: raw recordings of the electrocorticogram (ECoG), electromyogram (EMG), Hipp field activity, and neuronal discharge (unit) during W, SWS, and PS. Digitized units and acoustic stimuli are shown below. The left corner inset exhibits the post-stimulus time histogram (PSTH) in response to a pure tone-burst stimulus at the unit characteristic frequency. Bottom: the cross-correlation between Hipp field activity and auditory units firing was obtained by spike-triggered averaging. The insets show the Hipp power spectra (FFT) with the theta range in black. In this example, the neuronal discharge is phase-locked with the Hipp theta rhythm during SWS and PS whereas no temporal correlation appears in a W epoch. Cal.: ECoG, 0.5 mV; EMG, 0.1 mV; Hipp, 0.5 mV; unit, 0.1 mV; time, 1 s (modified from Velluti and Pedemonte (2002)).

In our experimental paradigm we tested cortical auditory neurone response to artificial sounds, clicks, or pure tones at the unit characteristic frequency. Afterwards, the unit response to a pre-recorded guinea pig call, 700-ms duration, was analyzed. Besides, the natural sound stimulus was delivered in a direct or inverted in time manner, evoking a different single unit response on passing from a waking state to SWS (Fig. 5.15), i.e., these cortical auditory neurones (and their neuronal network) can distinguish the differences even in sleep at the cortical as well as at the inferior colliculus level.

It is not known whether the lack of inferior colliculus unit activity during self-produced vocalization is due to a direct inhibitory input or the result of an inhibition on a second structure normally providing the external nucleus with auditory input. Anatomical studies have shown that the external nucleus is reciprocally connected with the periaqueductal gray (Radmilovich et al., 1991), a structure considered to be an important vocalization control area. The existence of such a direct connection suggests that the periaqueductal vocalization area exerts a gating function on the auditory input reaching the external nucleus of the inferior colliculus. This function might serve to reduce awareness of self-produced vocalizations in relation to, e.g., conspecific calls and other external stimuli. Neurones that are activated by external sounds, but not by self-produced vocalizations, have been found also in the auditory cortex. The fact that such neurones exist in the inferior colliculus suggests that selective attention processes take place already at midbrain level (Pérez-Perera et al., 2001; Pérez-Perera, 2002; Tammer et al., 2004; Bentancor et al., 2006).

When guinea pigs were stimulated with their own call, a "whistle" of 700-ms duration, the response of auditory cortex neurones was different in W and sleep. Furthermore, when the natural call was played backwards, i.e., inverted in time, the neuronal firing changed in W as well as during SWS. As depicted in Figure 5.15, the response to a "whistle" in W

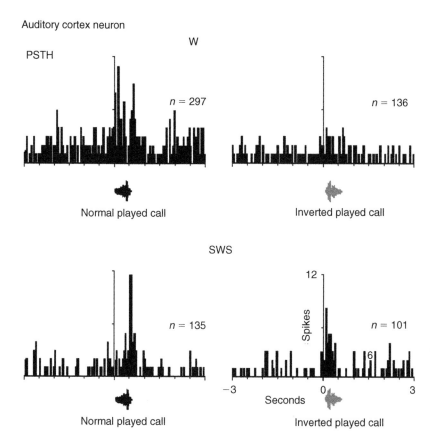

FIGURE 5.15 Response of an auditory cortex (AI) neurone evoked by a guinea pig pre-recorded natural call ("whistle"). It was played back in its natural mode, direct, and reversed in time, during W and SWS. The peristimulus time histogram (PSTH) during a W epoch exhibits the unit firing to the direct natural call decreasing by half when the same sound was played backwards, with an associated change in the pattern. During SWS the unit also exhibited firing shifts, depending on whether the sound was played normally or inverted in time. Thus, the units – included in some neuronal network – may recognize the difference between direct and inverted in time stimuli even in sleep. The response to the stimulus that is inverted in time in SWS is smaller and has shorter latency. A relation to the large part of the stimulus appears as possible (modified from Pérez-Perera et al. (2001); Pérez-Perera (2002)).

Auditory inferior colliculus neurone

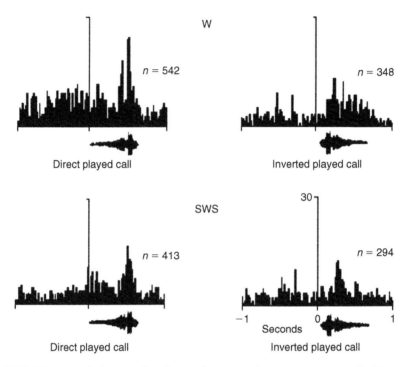

FIGURE 5.16 Inferior colliculus auditory single unit activity evoked by a guinea pig natural call (whistle). The stimuli were delivered directly or inverted in time that is played backwards. The natural stimulus was presented during W and SWS. As happens in a similar cortical recording, the inferior colliculus neurone exhibited firing number as well as pattern differences between the W and the sleep recordings. When the sound stimuli were played backwards, inverted in time, the auditory responses were present, although with differences, during W and also during SWS, i.e., the latter behavioral condition can also recognize the sound differences (modified from Bentancor et al. (2006)).

decreased when the same vocalization/call was presented reversed in time.

The same cortical auditory neurone recorded during an SWS epoch exhibited a totally different evoked response. The PSTH generated by normally played stimulus showed a peak in close temporal correlation with the "whistle" higher amplitude. On the other hand, when the natural call was played inverted in time, the evoked unit firing and latency decreased. Perhaps, the PSTH unit discharges peak was relevant in relation to the highest amplitude of the stimulus; in any case, it is evident that the cortical processing of complex sounds continues to be carried out in SWS.

The reported relationship between the auditory neurones and the hippocampal theta waves, phase-locking, was also present when the natural call stimulation was used. The cortical neurones studied exhibited cross-correlation (phase-locking) during W, SWS, and PS (Pérez-Perera et al., 2001; Pérez-Perera, 2002).

Inferior colliculus neurones have also shown a differential response to natural sound stimulation when presented direct or inverted in time (Fig. 5.16). Two factors may be involved in the response changes to direct or inverted sound stimulation: (a) a different temporal organization of the stimuli frequencies is present when presented backwards and (b) it may be also possible that the differences express a new "significance." Anyway, the unit and its associated neuronal network can distinguish both stimuli even in SWS (Bentancor et al., 2006).

A similar approach in birds also introduced auditory units firing changes during SWS, stimulating with the bird own song. The bird song system has neurones that respond to auditory stimuli, particularly to the bird's own song. Recordings at the high vocal center revealed that the single unit response to auditory stimuli is great during SWS monitored by electroencephalography (EEG) recordings. Spontaneous waking causes the ending of this firing within milliseconds. It is likely that if the vocal control system replays the song, "… then the replay functions to maintain the memory of the song in the adult" (Nick and Konishi, 2001).

Acoustic communication in noise

Humans are remarkably adept at acoustic communication in high levels of background noise (the cocktail-party effect). For efficient communication the average speech level should exceed that of the noise by 6 dB. But speech may be intelligible at negative signal-to-noise ratios for continuous speech in which the listener is familiar with the subject matter or if the speech and the noise are separated in space (known as spatial release from masking).

A wide variety of animals have also evolved behaviors that facilitate the ability to communicate in high levels of background noise. Both Old World (Sinnott et al., 1975) and New World (Brumm et al., 2004) monkeys increase the intensity of their vocalizations in a high-noise background (Lombard effect). In response to high-level ambient noise, frogs are able to adjust their calling pattern to minimize acoustic interference. In one experiment in which high-level periodic tones were broadcast for 90% of the 2.5 seconds stimulus period, male frogs in their natural habitat were able to initiate more calls in the 10% silent period than would be expected by chance alone (Zelick and Narins, 1983). Remarkably, these results persisted for aperiodic interfering tones, suggesting that the cue for call initiation is not entrainment to the stimulus rhythm but rather the cessation of interference, or the rapid reduction of ambient sound level. This was directly tested by broadcasting high-level interfering tones to calling male frogs that suddenly dropped in level. Fifty-nine percent of the males tested were able to initiate significantly more calls during a level drop of 4–6 dB than would be expected by chance.

Recently, several species of ranid frogs have been shown to produce a wide variety of vocalizations containing significant ultrasonic (>20 kHz) harmonics (Narins et al., 2004, Feng et al., 2006). In two species of *Odorrana* (Asian torrent frogs),

auditory evoked potentials and single unit recordings from
the inferior colliculus confirm auditory sensitivity up to
34 kHz. It is believed that ultrasonic sensitivity in these animals has evolved in response to selection pressure from the
high-level broadband ambient noise produced by the rushing
streams in these animals' environments.

Peter M. Narins
University of California
California, Los Angeles, USA

Conclusions

The three experimental approaches shown – the auditory neurone's firing rate, the discharge pattern, and the temporal correlation with the hippocampus theta rhythm – represent evidences of
sensory processing aspects that occur in sleep and waking. This
is also giving insight into how sensory information processing
and sleep physiology reciprocally affect each other, participating
in the processing and/or in the postulated active promotion of
sleep functions.

The changes in neuronal discharge rate and pattern in
response to constant stimuli indicate that the CNS modulates
the incoming auditory information according to the behavioral
state, from the auditory nerve to the auditory cortex. Likewise,
somatosensory (Pompeiano, 1970; Soja et al., 1998) and visual
neurones (Livingstone and Hubel, 1981; McCarley et al., 1983;
Gambini et al., 2002) exhibit changes in their firing rates in correlation with stages of sleep and W (review, Velluti, 1997). This is
consistent with the hypothesis that a general shift in the neuronal network/cell assembly's organization is involved in sensory
processing that occurs during sleep. This assumption is supported by magnetoencephalographic (MEG) study of auditory

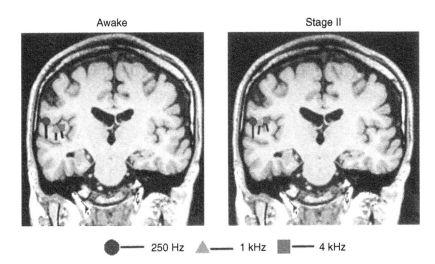

FIGURE 5.17   *Planum temporale* auditory cortical location of the M100 MEG component observed in response to three different sound frequency stimuli (250 Hz, 1 and 4 kHz) recorded in W and stage II sleep in human. The magnetometer was placed on the left hemisphere (C3 position) and the signal source was estimated using an equivalent current dipole (ECD) model. ECD overlapped on magnetic resonance imaging. The ECDs (dipoles) were localized deeper in response to the higher-frequency tones to the lower-frequency tones, while the three changed position on the cortex. The relatively great shifts in the cortical space exhibited by the dipoles demonstrate that the working network changed, surely including new cells elements and communications (modified from Naka et al. (1999)). See Plate 5.17 for the colour version of this figure.

stimulation during sleep performed in humans. It was observed that the dipole location changed in the auditory cortex on passing to SWS (Fig. 5.17), thus demonstrating the existence of a functional/anatomical network/cell assembly shift in the *planum temporale* area upon passing to sleep (Naka et al., 1999; Kakigi et al., 2003).

A number of neurones at different auditory *loci*, from the brainstem to the cortex itself, exhibited significant quantitative and qualitative changes in their evoked firing rate and pattern of discharge on passing to sleep. Most important, no neurone belonging to any auditory pathway level or cortex was observed to stop firing during sleep. In addition, our results indicate that the responsiveness of the auditory system during sleep can be considered preserved. Those neurones that continue to fire during sleep equal to their firing during W (~50% at the primary cortex) are probably related to a continuous monitoring of the environment, whereas the units that increase or decrease their evoked discharge would participate in sleep-related functions, probably associated with different sleep-related active neuronal networks. I cannot advance what their involvement in sleep neurophysiology could be, but it is my hypothesis that they are actively involved in sleep processes, e.g., sleep organization and maintenance, dreams content and mechanisms of arousal.

Functional magnetic resonance imaging (fMRI) and the evoked potentials cortical late waves in humans have also provided evidences that the sleeping brain can process auditory stimuli and detect meaningful events (see Chapter 2).

The temporal correlation between hippocampal theta rhythm and the firing of sensory units was shown at several levels, in the auditory pathway (Velluti and Pedemonte, 2002) and in visual neurones of the thalamus (Gambini et al., 2002). At a neuronal population scale, this phase-locking may result in a composite final signal that could be used in processes like attention, movement, and, in particular, auditory sensory incoming information processing. Furthermore, I am proposing that the phase-locking to the hippocampus theta waves adds a temporal dimension to the sensory processing, perhaps necessary for time-related perception. Every auditory stimulus, every sound, develops in time, that is why the CNS must have a functional possibility to encode this parameter. The hippocampus theta rhythm, being one of the most regular brain-generated low-frequency rhythms, may participate as an internal low-frequency clock acting as a time giver (Pedemonte et al., 1996b; Velluti et al., 2000; Pedemonte et al., 2001; Velluti and Pedemonte, 2002; Pedemonte and Velluti, 2005b).

Furthermore, discrimination of significant auditory signals from a background noise is a result of the enhancing excitatory and inhibitory periods in the unit responses to the acoustic stimulus under hippocampus theta influences (Parmeggiani et al., 1982). The phase precession of a CA1 cell and theta waves that occur when a rat is approaching a specific place – i.e., the unit appears earlier in relation to theta waves – constitutes an example of a temporal coding in the mammalian (Magee, 2003). This precession phenomenon has been also associated with non-spatial behaviors such as during PS (Buzsaki, 2002).

The temporal relationship between the neuronal firing and the hippocampus field activity is a varying phenomenon in the time domain that may be dependent on the interaction of a set of signals: (i) the hippocampal rhythm amplitude and/or frequency; (ii) the current state of the brain, awake or asleep; and (iii) the characteristics of the incoming sensory information.

The response of auditory neurones to the animal own vocalizations supports the experimental results obtained by using artificial sound (tone-bursts). In general, the population of cortical neurones stimulated with natural sounds showed W and SWS firing shifts and hippocampal theta phase-locking as in response to artificial stimuli. Furthermore, during SWS and using natural call stimulus, both responses were present although with some differences, perhaps representing a component of another processing category in a different neuronal network/cell assembly. In addition, it has been suggested the AI cortical region might serve a general purpose hub of the auditory pathway for the representation of complex sound features to be later complemented with higher auditory centers that further process high-level properties (Griffiths et al., 2004).

It is further suggested that the activity-dependent brain development during early life may not only occur during W (Marks et al., 1995; Roffwarg et al., 1966) but also during the long physiological sleep periods of newborns and infants. During early ontogenetic development, and maybe also in adulthood, the sensory information that reaches the CNS during sleep may "sculpt" the brain and participate in the adaptation to novel conditions.

The initial step toward an auditory learning process is the demonstration that the incoming corresponding information may be differentially processed in sleep. Once the auditory information is processed could be ready for learning, perhaps even in sleep. This is consistent with reports of learning during sleep in human newborns (Cheour et al., 2002), consolidation of perceptual learning of spoken language in sleep (Fenn et al., 2003), and visual discrimination improved after sleep (Stickgold et al., 2000a) observed in different sensory systems.

6

Auditory influences on sleep

"... sleep appears to be a behavioural state resulting from dynamic interactions of different physiologic functions in response to several endogenous (feeding, fatigue, temperature, instinctive drives) and exogenous (light–dark, temperature, food, season, social drives) cues. From the viewpoint of its determination, the mechanism appears so complex as to justify a theoretical distinction between *proximate, intermediate* and *remote* aspects of determination of sleep behaviour. This gradual approach to sleep behaviour avoids extending the category of rigid causal determination beyond the molecular and cellular levels and forcing experimental results to fit a reductionistic theory in spite of the fact that many elementary physiologic events character-ising sleep behaviour are not specific to sleep alone. In other words, sleep, like wakefulness, is a function of other interactive functions and not the unique result of the compelling influence of a segregated and highly specific neuronal network of the central nervous system".

(Parmeggiani, 2005)

The constantly present sensory incoming signals imposing conditions to the brain activity and, *vice versa*, the state of the brain imposing rules to the incoming information, is now the main objective (Fig. 6.1). The efferent pathways are the channels

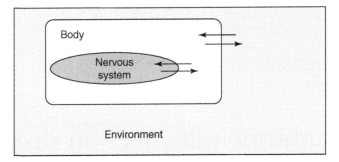

FIGURE 6.1   The nervous system placed inside the body has communicating channels that connect it with the environment and with the body itself through a diverse set of receptors.

connecting the central nervous system (CNS) with receptors and nuclei, e.g., the auditory efferent system arriving at the cochlea and nuclei (Chapter 1). Furthermore, in the visual system sleep also modifies the rat electroretinogram (Galambos et al., 1994) and exerts actions on ganglion cells (Cervetto et al., 1976).

In spite of the many membrane or circuital oscillators described, my view stresses the capacity of the sensory incoming information, from the outer and the inner (body) world, acting as a partial although relevant generator of the CNS basic activity, which performs continuously during wakefulness as well as sleep. It is my view that the sensory input is a decisive physiological fact that, acting on different brain areas, may support the basic brain activity also during sleep. Wakefulness or sleep characteristics have to be modulated on this continuous incoming flow of information (Fig. 6.2).

When some still unknown signals indicate the system to change the neural networks in order to enter the "sleeping mode," the sensory systems change along, shifting its wakefulness control characteristics into a "sensory sleeping mode"

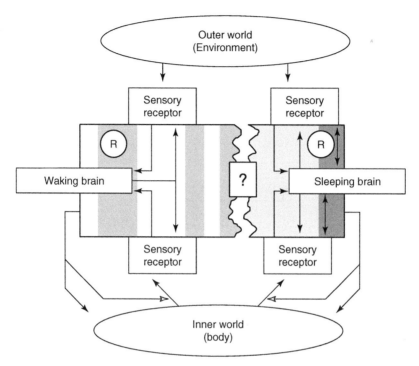

FIGURE 6.2   The waking and sleeping brain receives – during its different phases – sensory information from the outer world and the inner (body) world; this means there is an enormous and continuous barrage of incoming activity impinging on the brain, activating and/or deactivating diverse regions of the CNS. All the receptors are controlled by efferent connections and CNS actions are exerted on the body physiology both during wakefulness and sleep. The inner world receptors carry information from viscera, muscles, tendons and joints, from pressure and chemoreceptors, blood flowing noise, etc. There are also intra-cerebral receptors for blood sugar levels, hormones, etc.

function, thus producing a different processing form, we have to admit, according to sleep sensory-related needs:

- initiate and maintain sleep,
- keep a relative isolation from the environment,

- monitor the environment,
- produce wakefulness if the incoming information is significant or of high intensity,
- cooperate with the dreaming process and/or content,
- process memory-stored data,
- store new or processed data in memory.

Sensory input and sleep

The analysis of sensory functions during the sleep–waking cycle leads to the conclusion that normal sleep depends in many ways on the sensory input. It is suggested that the sleep and waking control networks are modulated by several inputs, and therefore, a proportion of "passive" effects must be associated with active functions for entering and maintaining normal sleep.

In general, the sensory input is a relevant signal. The total amount of sleep increments is produced under some experimental conditions facing sensory shifts (Velluti, 1997):

- Continuous somatosensory stimulation induces electroencephalography (EEG) synchronization and sleep.
- Total darkness increases sleep although just during a few days.
- Total silence, after bilateral cochlear destruction, increases the amount of sleep (Pedemonte et al., 1996b; Cutrera et al., 2000).
- Use of intracochlear implants in deaf humans, which shows sleep stages percentages to be different when compared with same patients before recovering hearing (Velluti et al., 2003).
- Lack of olfactory input after epithelial receptors lesion, which evokes a different proportion (%) of sleep stages and waking (López et al., 2007).

Partial increments in the occurrence of specific sleep stages are observed (review, Velluti, 1997):

- when rats are stimulated with sounds during any sleep stage;
- during stimulation with bright light, which produces slow wave sleep (SWS) increase in humans;
- during electrical stimulation of the olfactory bulb, which increases SWS in cats.

The sensory influences on sleep, such as the abolition or decrement of a sleep sign or stage (review, Velluti, 1997), are produced by:

- continuous light stimulation in rats that decreases paradoxical sleep (PS) for ~20 days;
- bilateral lesions of some vestibular nuclei that abolish rapid eye movements during PS up to 36 days;
- a long exposure to cold that produces decrement of PS leading to its deprivation;
- olfactory bulbectomy which decreases PS episodes incidence and its total amount for up to 15 days.

The lack of sensory inputs as well as their enhancement can produce sleep–waking imbalances, augmenting or diminishing sleep stages proportions, and provoking electrophysiological shifts. Thus, the induced changes in the waking and sleep networks lead to imbalances not simply explained by passive sleep production but also by introducing sensory sleep–active influences. A diversity of approaches supports this notion:

1. Sleep and sound are closely related. Environmental noises as well as regular, monotonous auditory stimuli, e.g., mother lullaby, are influences impeding or facilitating sleep.
2. The CNS and auditory system bioelectrical field activity – evoked potentials, magnetoencephalographic (MEG) evoked

activity – shown from early electrophysiological studies, vary in close correlation with wakefulness epochs and specially during sleep stages.

3. The auditory system single neuronal firing exhibits a variety of changes along the different nuclei and primary cortical *loci* linked to the sleep–wakefulness cycle in many ways: (a) increasing or decreasing firing on passing to sleep, (b) firing during sleep as during wakefulness, (c) changing the discharge pattern, (d) exhibiting theta rhythm phase-locking, with no auditory neurone stopping its firing on passing to sleep. Furthermore, Edeline et al. (2001) reported changes in the receptive field of cortical auditory neurones.

Therefore, it can be concluded that, when asleep, many auditory units are active, probably in association with diverse sleep-relevant cell assemblies. Moreover, when functionally shifting into a different neuronal network/cell assembly, a unit may contribute to the sleep process just by increasing, decreasing, or showing no firing shift, according to the new role in the new cell assembly association.

4. A MEG approach described anatomical place shifts of the sound evoked dipole in the human primary auditory cortex (*planum temporale*) on passing from wakefulness to sleep stage II (Naka et al., 1999; Kakigi et al., 2003). The dipole anatomical position shift obtained with MEG in the *planum temporale* is indicative of a change to a new neuronal group, a different neuronal network, already supported by single-unit studies (see Chapter 5). The MEG auditory evoked activity during sleep – its dipole – appears in a different cortical region from that during wakefulness, thus suggesting a new cell assembly/neuronal network participation.

5. The functional magnetic resonance imaging (fMRI), when combined with EEG recording, showed that the auditory stimuli produce bilateral activation in the human auditory cortex, parietal and frontal areas, both during wakefulness and sleep (Portas et al., 2000).

6. The positron emission tomography (PET) exhibits the cerebral blood flow (CBF) measurements in waking and sleep stages (Maquet et al., 2005). It is assumed that shifts in neuronal firing – glia and neurone activity – and oxygen consumption of the tissue induce proportional changes in local CBF, flow-activity coupling. An example of such possibility is the oxygen availability recording using an oxygen cathode (Velluti et al., 1965). When recording from a cat visual cortex the oxygen availability rises when flashes of light are presented to the animal (Fig. 6.3). The lower record shows the oxygen cathode response capacity to a changing air O_2 concentration, i.e., when the animal breathes 100% oxygen.

The data exhibited by fMRI strongly support the notion that the sleeping brain is able to process information, detecting meaningful events, as it can be observed in the unitary response in guinea pigs when a complex stimulus (the animal call) is played normally or in reverse (Chapter 5). A strong interaction between auditory input and sleep processes is put forward.

Some sleep researchers are, unconsciously, looking for a "sleep center" that does not exist. A CNS center may be real and useful for controlling functions such as the cardiovascular, the respiratory, etc., while on the other hand, sleep is not a function but a complete different CNS state. This means different brains for the diverse wakefulness conditions, for sleep stages I, II, SWS, and for PS with or without phasic components. Hence, sleep means a whole change of networks/cell assemblies, a new cooperative interaction among them, considering that a single network may subserve several different functions. The auditory dipole position change on passing to sleep appears as an advance, an experimental evidence, supporting the notion of a new configuration of neuronal networks all over the brain appearing when asleep.

A simple mollusc brain may serve as an example of functionally changing networks associated with different behaviors. The

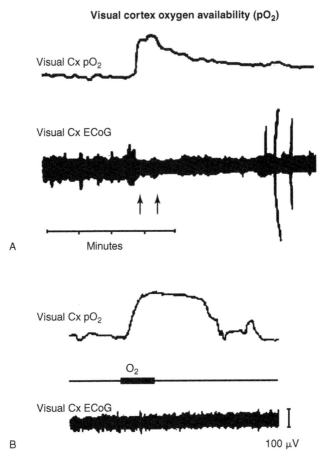

Visual cortex oxygen availability (pO₂)

FIGURE 6.3   Cortical oxygen availability (pO_2) during visual sensory stimulation in an awake cat. (A) An increase in the oxygen availability appears after light flashes (20/s) are presented for a few seconds to the animal's eyes with dilated pupils. (B) The oxygen cathode response capacity when the animal breathes pure oxygen is shown. This technique does not permit to calibrate the cathode because the diffusion coefficient at the electrode tip is never known (modified from Velluti et al. (1965)).

imbalance introduced by cell 1 activity in the *Tritonia* uncomplicated brain produces two functional networks related to two different movements (see Fig. 3.3).

In a complex brain such an imbalance may be carried out by a summation of physiological factors that, after some unknown signal/s – perhaps decreasing light intensity – produce sleep, i.e., a group of neuronal networks/cell assemblies progressively begin to prepare the brain to enter sleep. Partially supporting this assumption is the observation that when a human or animal is passing into sleep, the many variables usually recorded never occur in synchrony, but appear with seconds of difference among them, e.g., EEG slow activity, electromyogram decrement, eye movements, hippocampal theta rhythm frequency and amplitude, heart rate shifts, arterial pressure changes, breathing rhythm alterations, and so on.

Human and animal experimental data: effects of sound stimulation and auditory deprivation on sleep

The organization of human sleep is extremely sensitive to acoustic stimuli (Croome, 1977) and noise generally exerts an arousing influence on it (Muzet and Naitoh, 1977). A noisy nighttime ambiance leads to a decrease in total sleep time, particularly that of delta wave sleep (stage IV) and PS, with the consequent increase in the time spent in stage II and wakefulness (Vallet and Mouret, 1984; Terzano et al., 1990). Moreover, the remarkable sleep improvement after noise abatement (Vallet, 1982) suggests that the environment is being continuously scanned by the auditory system. This notion is also supported by results from single-unit recordings in sleeping animals previously described (Chapter 5).

Conflicting results have been reported concerning sound stimulation effects on sleep in animals. Using continuous high-intensity white-noise stimulation resulted in almost complete deprivation of PS in rabbits (Khazan and Sawyer, 1963), while in rats the same stimulation led to PS reduction without a decrease in the amount of SWS (van Twyver et al., 1966). Upon intense auditory stimulation, specifically during PS, the number of episodes increased without changing its total amount (Drucker-Colín et al., 1990), while the number of ponto-geniculate-occipital waves was enhanced in cats also stimulated with high intensity sound during the same sleep stage (Drucker-Colín et al., 1983; Ball et al., 1989). Moreover, auditory stimulation carried out recently in rats led to the conclusion that the pattern of PS occurrence was affected by stimulating both during PS and SWS, while the total amount of PS was substantially preserved (Amici et al., 2000; 2001). An animal study carried out in rats reported that the SWS-2, with high delta wave power, showed the highest arousal threshold when using a non-meaningful sound (Neckelmann and Ursin, 1993).

Auditory stimulation in animals as well as in humans seems to produce general actions on sleep processes, still not clearly defined, because stimulation during both sleep stages results in similar changes in the general sleep pattern.

Very interesting results, although old, reported the responsiveness of a conditioned cat to a 5-kHz "positive" tone (reinforced with classical and instrumental conditioning) during wakefulness (Buendía et al., 1963). It was orderly in wakefulness but different in each sleep phase, measured by the tone capacity to awaken the animal. During SWS the effectiveness of the 5-kHz tone for awakening the animal was not significantly changed, while the surrounding tones' ("negatives," never reinforced) capacity decreased. The tone separation ability of the system became sharper, indicative of a more precise discriminating mechanism during SWS; during PS the cats were practically non-responsive.

A question arises immediately: is there any relationship between this discriminative ability during SWS and the enhanced amplitude of the auditory nerve compound action potential and the increased amplitude of the auditory evoked potentials in all the brain *loci* studied up to the cortex itself, during this sleep phase (Chapter 1)? Moreover, is there any relationship with the increasing auditory single-unit firing in SWS? (Chapter 5).

On the other hand, total auditory deprivation in guinea pigs by surgical removal of both cochleae enhances SWS and PS by a similar proportion while reducing wakefulness, for up to 45 days post-lesion (Pedemonte et al., 1996b). The SWS and PS increments cited were determined mainly by an increase in the number of episodes with no change in single episode duration. The authors contend that the relative isolation from the outside world may elicit part of the change observed in deaf guinea pigs. Thus, eliminating an input to a complex set of networks, as the ones that regulate the sleep–waking cycle, would introduce functional shifts particularly if such input has some significance, as appears to be the case for the behavior under study: wakefulness and sleep.

Contrasting results have been reported although all of them are indicative of a correlation between both auditory input and the sleep–waking processes, supported by stimulation experiments as well as by the opposite approach, i.e., recording from deaf guinea pigs. More experimental data are clearly needed in the particular field of auditory sensory processes and sleep.

The many technical approaches reviewed (Fig. 6.4) support the notion that the sensory information in general and the auditory incoming information in particular exert influences on sleep through a dynamic neuronal participation in different sleep-related cell assemblies. Sector 7 in Figure 6.4 shows the sleep pattern of a deaf patient that has an intra-cochlear implant. This patient could be studied as a deaf person or as an almost normal-hearing person. When the sleep was analyzed during two nights at the sleep laboratory as deaf patient – intra-cochlear implant off – the sleep architecture and stages percentages were similar to the normal controls of a similar age. When performing the two-night recording as an almost normal-hearing person – implant on – the results were totally different with huge changes in the stages percentages. This result shows quite clearly the auditory influence on sleep already observed in deaf guinea pigs.

Finally, I have previously postulated that the auditory neurones firing in sleep at the same rate and pattern as during

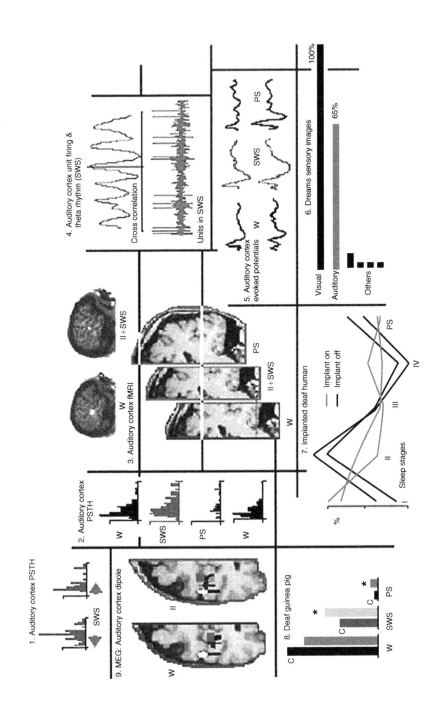

1. Auditory cortex PSTH

SWS

2. Auditory cortex PSTH

W

SWS

PS

W

3. Auditory cortex fMRI

W

II+SWS

4. Auditory cortex unit firing & theta rhythm (SWS)

Cross correlation

Units in SWS

5. Auditory cortex evoked potentials

W

SWS

PS

6. Dreams sensory images

Visual

Auditory

Others

100%

65%

7. Implanted deaf human

Implant on

Implant off

PS

IV

III

II

I

Sleep stages

%

8. Deaf guinea pig

C

C

*

C

*

W

SWS

PS

9. MEG: Auditory cortex dipole

W

II

wakefulness are those neurones that monitor the environment. These cell types increase their percentages from the brainstem up to the auditory primary cortical level. At the brainstem the units that shift their firing percentages perhaps are participating more closely in sleep-related regions. In the end, the units that increase/decrease their firing are postulated to be sleep-related active neurones, at cortical as well as at brainstem levels.

FIGURE 6.4   Diverse technical approaches supporting the postulated notion of the importance and possible active participation of the auditory input on sleep processes. Three human half-brain tomographic cuts (center) represent the three main functional possibilities: W, wakefulness; SWS, slow wave sleep; and PS, paradoxical sleep. (1) Post-stimulus time histogram (PSTH) changes of a cortical auditory neurone firing shift when stimulated with natural sound played directly or backward (Pérez-Perera et al., 2001). (2) PSTH of a cortical unit on passing from wakefulness to SWS and PS exhibits firing and pattern shifts (Peña et al., 1999). (3) Human auditory cortical imaging (fMRI) demonstrates activity during sleep (modified from Portas et al. (2000)). (4) The cortical auditory neurones can be phase-locked to hippocampal theta rhythm (Pedemonte et al., 2001). (5) Rat auditory cortical evoked potentials through the sleep–waking cycle show amplitude changes (Hall and Borbély, 1970). (6) The dream auditory "images" are present in 65% of dream recalls (McCarley and Hoffman, 1981). Human and guinea pig deafness influence sleep: (7) the human recorded with the intra-cochlear implant off and on shows different sleep stages percentages while (8) the guinea pig exhibits (bars) an increase in sleep time with decreasing wakefulness (Pedemonte et al., 1996b; Velluti et al., 2003). (9) The human MEG shows a place shift of the dipole evoked by three sound stimulating frequencies on passing to sleep stage II, demonstrating a change of neuronal network/cell assembly (modified from Kakigi et al. (2003)). See Plate 6.4 for the colour version of this figure.

Conclusions

Arriving at the end some memories about the beginnings come to me, about the first thought devoted to the subject of this book. Long ago, at the commencement of my research career, when looking at some newly collected auditory data at an oscilloscope, I asked myself where the observed changes were originated in the brain. Besides, what might happen to the data if the brain actually shifted state? Thus, after many experimental approaches and a lot of readings, from the auditory and sleep perspectives, the plan to analyze together both topics, became real. The first time I engaged in the study of auditory signals during sleep was in the late 1960s, although it came out as a first publication in 1989.

Complex patterns of activity are continuously generated in the brain which are spatially and temporally structured although remaining receptive to any sensory input. These are physiological conditions not usually taken into account when, e.g., modeling networks. The relevance of the sensory input is not considered at the level of significance it really has; the sleep passive theory expressed that the brain slept when the sensory input diminished or disappeared. Now it can be experimentally supported that a constant input is modulating the waking and sleep brain activity.

Of the billions of neurons that are active during waking life, I believe the same amount or so are also active in sleep. The brain does not stop its performance in sleep, although it changes all its characteristics. An intrinsic capacity of the human and animal brain, i.e., plasticity, is an evolutionary phenomenon that permitted to escape from the constraints of the genome becoming sensitive to environmental and internally generated inputs. Plasticity is part of the lifetime ongoing nervous system activity, even during sleep time.

The individual neuronal spikes can convey many bits of information about the incoming auditory stimulus and the specific and accurate discrimination of their meaning. It could in principle be based on the firing rate, firing number, firing pattern, and their relationship (phase-locking) to other general brain rhythm as the hippocampus theta waves postulated as a low-frequency timing code.

The results obtained through different experimental designs encourage me and my research team to critically study the possibility that auditory spikes entering the brain could play a role during sleep, perhaps not totally disconnected from the waking performance. This role may be related to both auditory information processing and sleep widely unknown processes. Moreover, the sleeping brain is a totally different one, i.e., sleep is another state that occupies about 30% of a human life. This simple number demonstrates the essential need for this physiological state, which must lead us to critically think and experimentally explore the brain changes that construct sleep.

Highlights from the auditory system standpoint

- All possibilities are present in the efferent system actions, i.e., the auditory nerve compound action potential, and the cochlear microphonic amplitude may increase and/or decrease depending on the way the system is activated (Table 1.2).
- No auditory neurone was ever recorded stopping firing on passing to sleep, neither in the sub-cortical nuclei nor in the primary cortex.
- The changes in neuronal discharge rate and pattern in response to constant stimuli indicate that the CNS modulates

the incoming auditory information according to the behavioral state.

- Deafness, human and experimental (guinea pigs), introduces changes in the sleep stages amount and in total sleep time.
- A new hypothesis regarding the efferent auditory system function is presented: the efferent activity is intended to establish a *synergy* between the auditory system and the ever-changing CNS status.
- During early ontogenetic development the auditory sensory information that reaches the CNS during sleep may "sculpt" the brain and participate in the adaptation to novel circumstances. The demonstration that the incoming auditory information may be differentially processed in sleep and thus, could be ready for learning perhaps even during sleep.

Highlights from the sleep state viewpoint

- Sleep is a CNS state during which different functions take place.
- The physiology of different systems is modulated by the wakefulness–sleep cycle: the processing of sensory information, the oneiric activity, the cardiovascular and respiratory functions, the endocrine functions, as well as the body temperature control, the homeostasis and the energetic metabolism, all happen to change depending on the moment of the cycle.
- A functional shift in the neuronal networks/cell assemblies constitutes the foundation of the changes when passing from wakefulness to sleep. This assumption is supported by the auditory unitary analysis and the place shift of the magnetoencephalographic auditory evoked dipole in the *planum temporale*, both studied during sleep and waking.

- We are able to perceive an auditory stimulus to process and compare it to information stored in our memory and make decisions such as waking up or continue sleeping. These results suggest that neurones participate in different neuronal networks which functionally contribute in multiple processings besides the specific sensory one.
- The complete absence of auditory input produces modifications in sleep and wakefulness, increasing both sleep phases to the detriment of wakefulness. The sleeping brain imposes conditions for the incoming and processing of auditory sensory information through its efferent system; furthermore, it is my tenet that what is observed for the auditory system can be valid for all the sensory systems.
- We are still far from knowing what processes are triggered in the CNS to achieve sleep. However, it seems essential for cognitive development while memory and learning appear as relevant functions performed during sleep.
- A proposed *psycho-physiological information homeostasis* maintains the memory load within a range that the brain needs to maintain in order to preserve the cognitive functions. To remember or to forget are part of our brain goals in order to embrace a healthy status.

This book is the most recent step toward a rather synthetic approach in trying to put together both the sleep state and the auditory information processing. A general look at both subjects and some new experimental and theoretical approaches have been introduced.

References

Adametz, J.H. (1959). Recovery of functions in cats with rostral lesions. *J. Neurosurg.*, 16: 85–97.

Adey, W.R., Dunlop, C.W. and Hendrix, C.E. (1960). Hippocampal slow waves distribution and phase relations in the course of approach learning. *Arch. Neurol.*, 3: 74–90.

Adey, W.R., Kado, R.T., McIlwain, J.T. and Walter, D.O. (1966). The role of neuronal elements in regional cerebral impedance changes in alerting, orienting and discriminative responses. *Exp. Neurol.*, 15: 490–510.

Affani, J.M. and Cervino, C.O. (2005). Interactions between sleep, wakefulness and the olfactory system. In: P.L. Parmeggiani and R.A. Velluti (Eds), *The Physiologic Nature of Sleep*. Imperial College Press, London.

Alain, C., Arnott, S.R., Hevenor, S., Graham, S. and Grady, C.L. (2001). "What" and "where" in the human auditory system. *Proc. Natl. Acad. Sci. USA*, 98: 12301–12306.

Amadeo, M. and Shagass, C. (1973). Brief latency click-evoked potentials during waking and sleep in man. *Psychophysiology*, 10: 244–250.

Amato, G., Lagrutta, V. and Enia, F. (1969). The control exerted by the auditory cortex on the activity of the medial geniculate body and inferior colliculus. *Arch. Sci. Biol. (Bologna)*, 53: 291–313.

Amici, R., Domeniconi, R., Jones, C.A., Morales-Cobas, G., Perez, E., Tavernese, L., Torterolo, P., Zamboni, G. and Parmeggiani, P.L. (2000). Changes in REM sleep occurrence due to rhythmical auditory stimulation in the rat. *Brain Res.*, 868: 241–250.

Amici, R., Morales-Cobas, G., Jones, C.A., Perez, E., Torterolo, P., Zamboni, G. and Parmeggiani, P.L. (2001). REM sleep enhancement due to rhythmical auditory stimulation in the rat. *Behav. Brain Res.*, 123: 155–163.

Amici, R., Jones, C.A., Perez, E. and Zamboni, G. (2005). A physiological view of REM sleep structure. In: P.L. Parmeggiani and R.A. Velluti (Eds), *The Physiologic Nature of Sleep*. Imperial College Press, London, pp. 161–185.

Andersson, J.L., Onoe, H., Hetta, J., Lidstrom, K., Valind, S., Lilja, A., Sundin, A., Fasth, K.J., Westerberg, G., Broman, J.E., Watanabe, Y. and Langstrom, B. (1998). Brain networks affected by synchronized sleep visualized by positron emission tomography. *J. Cereb. Blood Flow Metab.*, 18: 701–715.

Arbib, M.A. (2002). The Handbook of Brain Theory and Neural Networks, 2nd Ed., MIT Press, Cambridge, MA.

Aserinski, E. and Kleitman, N. (1953). Regularly occurring periods of eye motility and concomitant phenomena during sleep. *Science*, 118: 273–274.

Atienza, M. and Cantero, J.L. (2001). Complex sound processing during human REM sleep by recovering information from long-term memory as revealed by the mismatch negativity (MMN). *Brain Res.*, 901: 151–160.

Atienza, M., Cantero, J.L. and Gómez, C. (1997). The mismatch negativity component reveals the sensory memory during REM sleep in humans. *Neurosci. Lett.*, 237: 21–24.

Atienza, M., Cantero, J.L. and Escera, C. (2001). Auditory information processing during human sleep as revealed by event-related brain potentials. *Clin. Neurophysiol.*, 112: 2031–2045.

Atienza, M., Cantero, J.L. and Domínguez-Marín, E. (2002). Mismatch negativity (MMN): an objective measure of sensory memory and long-lasting memories during sleep. *Int. J. Psychophysiol.*, 46: 215–225.

Baghdoyan, H.A., Rodrigo-Angulo, M.L., McCarley, R.W. and Hobson, J.A. (1984). Site-specific enhancement and suppression of desynchronized sleep signs following cholinergic stimulation of three brainstem regions. *Brain Res.*, 306: 39–52.

Ball, W.A., Morrison, A.R. and Ross, R.J. (1989). The effects of tones on PGO waves in slow wave sleep and paradoxical sleep. *Exp. Neurol.*, 104: 251–256.

Basheer, R., Porkka-Heiskanen, T., Strecker, R.E., Thakkar, M.M. and McCarley, R.W. (2000). Adenosine as a biological signal mediating sleepiness following prolonged wakefulness. *Biol. Signals Recept.*, 9: 319–327.

Bastien, C., Croewley, K.E. and Colrain, I.M. (2002). Evoked potential components unique to non-REM sleep: relationship to evoked K-complexes and vertex sharp waves. *Int. J. Psychophysiol.*, 46: 257–274.

Bastuji, H. and García-Larrea, L. (1999). Evoked potentials as a tool for the investigation of human sleep. *Sleep Med. Rev.*, 3: 23–45.

Bastuji, H. and García-Larrea, L. (2005). Human auditory information processing during sleep. In: P.L. Parmeggiani and R.A. Velluti (Eds), *The Physiologic Nature of Sleep*. Imperial College Press, London, pp. 509–534.

Bastuji, H., García-Larrea, L., Bertrand, O. and Mauguière, F. (1988). BAEP latency changes during nocturnal sleep are not correlated with sleep stages but with body temperature variations. *Electroencephgr. Clin. Neurophysiol.*, 70: 9–15.

Bastuji, H., García-Larrea, L., Franc, C. and Mauguière, F. (1990). Sleep related modifications of auditory cognitive potentials. A topographic study during physiological all-night sleep. *10th Congress of the European Sleep Research Society*, Strasbourg.

Bastuji, H., García-Larrea, L., Franc, C. and Mauguière, F. (1995). Brain processing of stimulus deviance during slow-wave and paradoxical sleep: a study of human auditory evoked responses using the oddball paradigm. *J. Clin. Neurophysiol.*, 12: 155–167.

Bastuji, H., Perrin, F. and García-Larrea, L. (2002). Semantic analysis of auditory input during sleep: studies with event related potentials. *Int. J. Psychophysiol.*, 46: 243–255.

Baudena, P., Halgren, E., Heit, G. and Clarke, J.M. (1995). Intracerebral potentials to rare target and distractor auditory and visual stimuli. III. Frontal cortex. *Electroencephgr. Clin. Neurophysiol.*, 94: 251–264.

Baust, W., Berlucchi, G. and Moruzzi, G. (1964). Changes in the auditory input in wakefulness and during the synchronized and desynchronized stages of sleep. *Arch. Ital. Biol.*, 102: 657–674.

Bentancor, C., Pedemonte, M. and Velluti. R.A. (2006). Actividad neuronal del colículo inferior durante el ciclo sueño-vigilia en respuesta a vocalizaciones de la especie. *Physiol. Mini-Rev.*, 2: 65.

Berland, I. and Pratt, H. (1995). P300 in response to the subject's own name. *Electroencephgr. Clin. Neurophysiol.*, 96: 472–474.

Berlucchi, G., Munson, J.B. and Rizzolatti, G. (1967). Changes in click evoked responses in the auditory system and the cerebellum of free-moving cats during sleep and waking. *Arch. Ital. Biol.*, 105: 118–135.

Bernal, B. and Altman, N. (2001). Auditory functional MR imaging. *Am. J. Roentgenol.*, 176: 1009–1015.

Best, P.J., White, A.M. and Minai, A. (2001). Spatial processing in the brain: the activity of hippocampal place cells. *Annu. Rev. Neurosci.*, 24: 459–486.

Born, A.P., Law, I., Lund, T.E., Rostrup, E., Hanson, L.G., Wildschiodtz, G., Lou, H.C. and Paulson, O.B. (2002). Cortical deactivation induced by visual stimulation in human slow-wave sleep. *Neuroimage*, 17: 1325–1335.

Bowker, R.M. and Morrison, A.R. (1976). The startle reflex and PGO spikes. *Brain Res.*, 102: 185–190.

Brandenberger, G. (1993). Episodic hormone release in relation to REM sleep. *J. Sleep Res.*, 2: 193–198.

Brandenberger, G. (2005). Endocrine correlates of sleep in humans. In: P.L. Parmeggiani and R.A. Velluti (Eds), *The Physiologic Nature of Sleep.* Imperial College Press, London, pp. 433–453.

Braun, A.R., Balkin, T.J., Wesensten, N.J., Carson, R.E., Varga, M., Baldwin, P., Selbie, S., Belenky, G. and Herscovititch, P. (1997). Regional cerebral blood flow throughout the sleep–wake cycle. An H2(15)O PET study. *Brain*, 120: 1173–1197.

Braun, A.R., Balkin, T.J., Wesesten, N.J., Gwadry, F., Carson, R.E., Varga, M., Baldwin, P., Belenky, G. and Herscovitch, P. (1998). Dissociated pattern of activity in visual cortices and their projections during human rapid eye movement sleep. *Science*, 279: 91–95.

Brázdil, M., Rektor, I., Dufek, M., Daniel, P., Jurák, P. and Kuba, R. (1999). The role of frontal and temporal lobes in visual discrimination task – depth ERP studies. *Neurophysiol. Clin.*, 29: 339–350.

Brázdil, M., Rektor, I., Daniel, P., Dufek, M. and Jurak, P. (2001). Intracerebral event-related potentials to subthreshold target stimuli. *Clin. Neurophysiol.*, 112: 650–661.

Brown, R.E. and McCarley, R.W. (2005). Neurotransmitters, neuromudulators, and sleep. In: P.L. Parmeggiani and R.A. Velluti (Eds), *The Physiologic Nature of Sleep*. Imperial College Press, London, pp. 45–75.

Bremer, F. (1935). Cerveau "isole" et physiologie du sommeil. *C.R. Soc. Biol.*, 118: 1235–1241.

Brownell, W.E., Bader, C.R., Bertrand, D. and de Ribaupierre, Y. (1985). Evoked mechanical response of isolated cochlear outer hair cell. *Science*, 277: 194–196.

Buendía, N., Sierra, G., Goode, M. and Segundo, J.P. (1963). Conditioned and discriminatory responses in wakeful and sleeping cats. *Electroencephgr. Clin. Neurophysiol. Suppl.*, 24: 199–218.

Buño, W. and Velluti, J.C. (1977). Relationship of hippocampal theta cycle with bar pressing during self-stimulation. *Physiol. Behav.*, 19: 615–621.

Buño, W., Velluti, R., Handler, P. and García-Austt, E. (1966). Neural control of the cochlear input in the wakeful free guinea pig. *Physiol. Behav.*, 1: 23–35.

Burgess, A.P. and Gruzelier, J.H. (1997). Short duration synchronization of human theta rhythm during recognition memory. *Neurol. Rep.*, 8: 1039–1042.

Buszaki, G. (1996). The hippocampus-neocortical dialogue. Cereb. Cortex, 6: 61–92.

Buszaki, G. (2002). Theta oscillations in the hippocampus. *Neuron*, 33: 325–340.

Caird, D. (1991). Processing in the colliculi. In: R.A. Altshuler, R.P. Bobbin, B.M. Clopton and D.W. Hoffman (Eds), *Neurobiology of Hearing: Central Auditory System*. Raven Press, New York, pp. 253–292.

Campbell, K. (1992). Evoked potentials measures of information processing during natural sleep. In: R. Broughton and R. Ogilvie (Eds), *Sleep, Arousal and Performance*. Birkhauser, Boston, pp. 88–116.

Campbell, K. and Bartoli, E. (1986). Human auditory evoked potentials during natural sleep. *Electroencephgr. Clin. Neurophysiol.*, 65: 142–149.

Campbell, K., Bell, I. and Bastien, C. (1992). Evoked potential measures of information processing during natural sleep. In: R.J. Broughton and R.D. Ogilvie (Eds), *Sleep, Arousal, and Performance*. Birkhauser, Boston/ Basel/Berlin, pp. 89–116.

Cannon, W.B. (1929). Organization for physiological homeostasis. *Physiol. Rev.*, 9: 399–431.

Cantero, J.L., Atienza, M., Stickgold, R., Kahana, M.J., Madsen, J.R. and Kocsis, B. (2003). Sleep-dependent θ oscillations in the human hippocampus and neocortex. *J. Neurosci.*, 23: 10893–10897.

Cardinali, D.P. (2005). The use of melatonin as a chronobiotic-cytoprotective agent in sleep disorders. In: P.L. Parmeggiani and R.A. Velluti (Eds), *The Physiologic Nature of Sleep*. Imperial College Press, London, pp. 455–488.

Cardinali, D.P. and Pevet, P. (1998). Basic aspects of melatonin action. *Sleep Med. Rev.*, 2: 175–190.

Carr, C.E. and Konishi, M. (1990). A circuit for detection of interaural time differences in the brain stem of the barn owl. *J. Neurosci.*, 10: 3227–3246.

Cartwright, R.D. (1974). The influence of a conscious wish on dreams: a methodological study of dream meaning and function. *J. Abnorm. Psychol.*, 83: 387–393.

Cazard, P. and Buser, P. (1963). Modification des réponses sensorielles corticales par stimulation de l'hippocampe dorsal chez le lapin. *Electroencephgr. Clin. Neurophysiol.*, 15: 413–425.

Cervetto, L., Marchiafava, P.L. and Pasino, E. (1976). Influence of efferent retinal fibers on responsiveness of ganglion cell to light. *Nature*, 269: 56–57.

Cespuglio, R., Colas, D. and Gautier-Sauvigné, S. (2005). Energy processes underlying the sleep–wake cycle. In: P.L. Parmeggiani and R.A. Velluti (Eds), *The Physiologic Nature of Sleep*. Imperial College Press, London, pp. 3–21.

Chemelli, R.M., Willie J.T., Sinton, C.M., Elmquist, J.K., Scammell, T., Lee, C., Richardson, J.A., Williams, S.C., Xiong, Y., Kisanuki, Y., Fitch, T.E., Nakazato, M., Hammer, R.E., Saper, C.B. and Yanagisawa, M. (1999). Narcolepsy in orexin knockout mice: molecular genetics of sleep regulation. *Cell*, 98: 437–451.

Cheour, M., Martynova, O., Naatanen, R., Erkkola, R., Sillanpaa, M., Kero, P., Raz, A., Kaipio, M.-L., Hiltunen, J., Aaltonen, O., Savela, J. and Hamalainen, H. (2002). Speech sounds learned by sleeping newborns. *Nature*, 415: 599–600.

Chow, K.L. and Randall, W. (1964). Learning and retention in cats with lesions in reticular formation. *Psychon. Sci.*, 1: 259–260.

Cicogna, P., Cavallero, C. and Bosinelli, M. (1991). Cognitive aspects of mental activity during sleep. *Am. J. Psychol.*, 104: 413–425.

Cipolli, C. (2005). Sleep and memory. In: P.L. Parmeggiani and R.A. Velluti (Eds), *The Physiologic Nature of Sleep*. Imperial College Press, London, pp. 601–629.

Clemente, C.D. and Sterman, M.B. (1963). Cortical synchronization and sleep patterns in acute restrained and chronic behaving cats induced by basal forebrain stimulation. *Electroencephgr. Clin. Neurophysiol.*, 24: 172–187.

Coenen, A.M. (1995). Neuronal activities underlying the electroencephalogram and evoked potentials of sleeping and waking: implications for information processing. *Neurosci. Biobehav. Rev.*, 19: 447–463.

Cordeau, J.P., Moreau, A., Beaulnes, A. and Laurin, C. (1963). EEG and behavioral changes following micro-injections of acetylcholine and adrenaline in the brainstem of the cats. *Arch. Ital. Biol.*, 101: 30–47.

Cote, K.A. (2002). Probing awareness during sleep with the auditory odd-ball paradigm. *Psychophysiology*, 46: 227–241.

Croome, D.J. (1977). Noise and sleep. *Noise, Building and People.* Pergamon Press, London, pp. 101–109.

Cutrera, R., Pedemonte, M., Vanini, G., Goldstein, N., Savorini, D., Cardinali, D.P. and Velluti, R.A. (2000). Auditory deprivation modifies biological rhythms in golden hamster. *Arch. Ital. Biol.*, 138: 285–293.

Czisch, M., Wetter, T.C., Kaufmann, C., Pollmacher, T., Holsboer, F. and Auer, D.P. (2002). Altered processing of acoustic stimuli during sleep: reduced auditory activation and visual deactivation detected by a combined fMRI/EEG study. *Neuroimage*, 16: 251–258.

Czisch, M., Wehrle, R., Kaufmann, C., Wetter, T.C., Holsboer, F., Pollmächer, T. and Auer, D.P. (2004). Functional MRI during sleep: BOLD signal decreases and their electrophysiological correlates. *Eur. J. Neurosci.*, 20: 566–574.

Datta, S. (1997). Cellular basis of pontine ponto-geniculo-occipital wave generation and modulation. *Cell. Mol. Neurobiol.*, 17: 341–365.

Dauvilliers, Y., Maret, S. and Tafti, M. (2005). Genetics of normal and pathological sleep in humans. *Sleep Med. Rev.*, 9: 91–100.

Davis, H. and Yoshie, N. (1963). Human evoked cortical responses to auditory stimuli. *Physiologist*, 6: 164.

Davis, H., Davis, P.A., Loomis, A.L., Harvey, E.N. and Hobart, G. (1939). Electrical reactions of the human brain to auditory stimulation during sleep. *J. Neurophysiol.*, 2: 500–514.

Deiber, M.P., Ibañez, V., Bastuji, H., Fischer, C. and Mauguière, F. (1989). Changes of middle latency auditory evoked potentials during natural sleep in humans. *Neurology*, 39: 806–813.

de Lugt, D.R., Loewy, D.H. and Campbell, K.B. (1996). The effect of sleep onset on event related potentials with rapid rates of stimulus presentation. *Electroencephgr. Clin. Neurophysiol.*, 6: 484–492.

Délano, P.H., Elgueda, D., Hamame, C.M. and Robles, L. (2007). Selective attention to visual stimuli reduces cochlear sensitivity in chinchillas. *J. Neurosci.*, 27: 4146–4153.

Delgado-García, J.M. and Gruart, A. (2002). The role of interpositus nucleus in eyelid conditioned responses. *Cerebellum* 1: 289–308.

Desmedt, J.E. (1975). Physiological studies of the efferent recurrent auditory system. In: W.D. Keidel and D. Neff (Eds), *Handbook of Sensory Physiology.* Springer, Berlin, pp. 219–246.

Doppelmayr, M., Klimesch, W., Schwaiger, J., Auinger, P. and Winkler, T. (1998). Theta synchronization in the human EEG and episodic retrieval. *Neurosci. Lett.*, 257: 41–44.

Drucker-Colín, R., Bernal-Pedraza, J., Fernández-Cancino, F. and Morrison, A.R. (1983). Increasing PGO spike density by auditory stimulation increases the duration and decreases the latency of rapid eye movements (REM) sleep. *Brain Res.*, 278: 308–312.

Drucker-Colín, R., Arankowsky-Sandoval, G., Próspéro-García, O., Jiménez-Anguiano, A. and Merchant, H. (1990). The regulation of REM sleep: some considerations on the role of vasoactive intestinal peptide, acetylcholine, and sensory modalities. In: M. Mancia and G. Marini (Eds), *The Diencephalon and Sleep*. Raven Press, New York, pp. 313–330.

Durmer, J.S. and Dinges, D.F. (2005). Neurocognitive consequences of sleep deprivation. In: K.L. Roos (Ed.), *Sleep in Neurological Practice. Semin. Neurobiol.*, 25: 117–129.

Edeline, J.-M. (2003). The thalamo-cortical auditory receptive fields: regulation by the sates of vigilance, learning and neuromodulatory systems. *Exp. Brain Res.*, 153: 554–572.

Edeline, J.-M. (2005). Learning-induced plasticity in the thalamo-cortical auditory system: Should we move from rate to temporal code descriptions? In: R. Konig, P. Heil, E. Budinger and H. Scheich (Eds), *The Auditory Cortex*. Lawrence Erlbaun Associates, Mahwah, NJ/London, pp. 365–382.

Edeline, J.-M., Manunta, Y. and Hennevin, E. (2000). Auditory thalamus neurones during sleep: changes in frequency selectivity, threshold, and receptive field size. *J. Neurophysiol.*, 84: 934–952.

Edeline, J.-M., Dutrieux, G., Manunta, G. and Hennevin, E. (2001). Diversity of receptive field changes in auditory cortex during natural sleep. *Eur. J. Neurosci.*, 14: 1865–1880.

Elgoyhen, A.B., Vetter, D.E., Katz, E., Rothlin, C.V., Heinemann, S.F. and Boulter, J. (2001). α10: a determinant of nicotinic cholinergic receptor function in mammalian vestibular and cochlear mechanosensory hair cells. *Proc. Natl. Acad. Sci. USA*, 98: 3501–3506.

Erwin, R. and Buchwald, J. (1986). Midlatency auditory evoked responses: differential effects of sleep in the human. *Electroencephgr. Clin. Neurophysiol.*, 65: 383–392.

Escera, C., Alho, K., Schroger, E. and Winkler, I. (2000). Involuntary attention and distractibility as evaluated with event-related brain potentials. *Audiol. Neurootol.*, 5: 151–166.

Esteban, S., Nicolau, M.C., Gamundi, A., Akaarir, M. and Rial, R. (2005). Animal sleep: philogenetic correlations. In: P.L. Parmeggiani and R.A. Velluti

(Eds), *The Physiologic Nature of Sleep*. Imperial College Press, London, pp. 207–245.

Evarts, E.V. (1964). Temporal pattern of discharge of pyramidal tract neurons during sleep and waking in the monkey. *J. Neurophysiol.*, 27: 152–171.

Evarts, E.V., Bental, E., Bihari, B. and Huttenlocher, P.R. (1962). Spontaneous discharge of single neurons during sleep and waking. *Science*, 135: 726–728.

Faingold, C.L., Gehlbach, G. and Caspary, D. (1991). Functional pharmacology of inferior colliculus neurons. In: R.A. Altshuler, R.P. Bobbin, B.M. Clopton and D.W. Hoffman (Eds), *Neurobiology of Hearing: Central Auditory System*. Raven Press, New York, pp. 223–251.

Faye-Lund, H. (1985). The neocortical projections to the inferior colliculus in the albino rat. *Anat. Embryol.*, 173: 53–70.

Faye-Lund, H. (1986). Projections from the inferior colliculus to the superior olivary complex in the albino rat. *Anat. Embryol.*, 175: 35–52.

Feliciano, M. and Potashner, S. (1995). Evidence for a glutamatergic pathway from the guinea-pig auditory cortex to the inferior colliculus. *J. Neurochem.*, 65: 1348–1357.

Fenn, K.M., Nusbaum, H.C. and Margoliash, D. (2003). Consolidation during sleep of perceptual learning of spoken language. *Nature*, 425: 614–616.

Fex, J. (1962a). Auditory activity in centrifugal and centripetal cochlear fibers in cat. A study of a feed-back system. *Acta Physiol. Scand. Suppl.*, 189: 1–98.

Fex, J. (1962b). Auditory activity in centrifugal and centripetal cochlear fibers in the cat. *Acta Physiol. Scand.*, 55: 2–68.

Fex, J. (1967). Efferent inhibition in the cochlea related to hair-cell activity: a study of postsynaptic activity of the crossed olvo-cochlear fibers in the cat. *J. Acoust. Soc. Am.*, 41: 666–675.

Foulkes, W.D. (1962). Dream reports from different stages of sleep. *J. Abnorm. Soc. Psychol.*, 65: 14–25.

Franzini, C. (1992). Brain metabolism and blood flow during sleep. *J. Sleep Res.*, 1: 3–36.

Froehlich, P., Collet, L., Valaxt, J.L. and Morgon, A. (1993). Sleep and active cochlear micromechanical properties in human subjects. *Hear. Res.*, 66: 1–7.

Galambos, R. (1956). Suppression of auditory nerve activity by stimulation of efferent fibers to cochlea. *J. Neurophysiol.*, 19: 424–437.

Galambos, R. and Velluti, R.A. (1968). Evoked resistance shifts in unanesthetized cats. *Exp. Neurol.*, 22: 243–252.

Galambos, R., Juhász, G., Kékesi, A.K., Nyitrai, G. and Szilágy, N. (1994). Natural sleep modifies the rat electroretinogram. *Proc. Natl. Acad. Sci. USA*, 91: 5153–5157.

Galkine, V.S. (1933). On the importance of the receptors for the working of the higher divisions of the nervous system. *Arkh. Biol. Nauk.*, 33: 27–55.

Gambini, J.P., Velluti, R.A. and Pedemonte, M. (2002). Hippocampal theta rhythm synchronized visual neurones in sleep and waking. *Brain Res.*, 926: 137–141.

García-Austt, E. (1963). Influence of the states of awareness upon sensory evoked potentials. *Electroencephgr. Clin. Neurophysiol. Suppl.*, 24, 76–89.

García-Austt, E. (1984). Hippocampal level of neural integration. In: E. Ajmone-Marsan and F. Reinoso-Suárez (Eds), *Cortical Integration Basic Archicortical and Cortical Association Levels of Neuronal Integrations*. IBRO Monograph Series, Raven Press, New York, pp. 91–104.

García-Austt, E., Velluti, R.A. and Villar, J.I. (1968). Changes in brain pO2 during paradoxical sleep in cats. *Physiol. Behav.*, 3: 477–485.

Garzón, M. (1996). Estudio morfofuncional de los núcleos reticular oral y reticular caudal del tegmento pontino como regiones generadoras de sueño paradójico. *Tesis Doctoral*, Universidad Autónoma de Madrid.

Gaztelu, J.M., Romero-Vives, M., Abraira, V. and García-Austt, E. (1994). Hippocampal EEG theta power density is similar during slow-wave sleep and paradoxical sleep. A long-term study in rats. *Neurosci. Lett.*, 172: 31–34.

George, R., Haslett, W.L. and Jenden, D.J. (1964). A cholinergic mechanism in the brainstem reticular formation: induction of paradoxical sleep. *Int. J. Neuropharmacol.*, 3: 541–552.

Getting, P.A. (1989). Emerging principles governing the operation of neural networks. *Annu. Rev. Neurosci.*, 12: 185–204.

Gillin, J.C., Sitaram, N., Janowsky, D., Risch, C., Huey, L. and Storch, F.I. (1985). In: A. Wauquier, J.M. Gaillard, J.M. Monti and M. Radulovacki (Eds), *Cholinetrgic Mechanisms in REM Sleep. Neurotransmitters and Neuromodulators*. Raven Press, New York, pp. 153–164.

Giuditta, A., Ambrosini, M.V., Montagnese, P., Mandile, P., Cotrugno, M., Grassi, Z.G. and Vescia, S. (1995). The sequential hypothesis of the function of sleep. *Behav. Brain Res.*, 69: 157–166.

Goldstein-Daruech, N., Pedemonte, M., Inderkum, A. and Velluti, R.A. (2002). Effects of excitatory amino acid antagonists on the activity of inferior colliculus neurons during sleep and wakefulness. *Hear. Res.*, 168: 174–180.

Grastyán, E., Lissák, K. and Madarász, I. (1959). Hippocampal activity during the development of conditioned reflex. *Electroencephgr. Clin. Neurophysiol.*, 11: 409–430.

Greenberg, J.H. (1980). Sleep and cerebral circulation. In: J. Orem and Ch.D. Barnes (Eds), *Physiology in Sleep*. Academic Press, New York, pp. 57–94.

Griffiths, T.D., Warren, J.D., Scott, S.K., Nelken, I. and King, A.J. (2004). Cortical processing of complex sound: a way forward? *Trends Neurosci.*, 27: 181–185.

Guillery, R.W., Feig, S.L. and Lazsádi (1998). Paying attention to the thalamic reticular nucleus. *Trends Neurosci.*, 21: 28–32.

Guinan, J.J. (1986). Effects of efferent neural activity on cochlear mechanics. *Scand. Audiol. Suppl.*, 25: 53–62.

Hagamen, W.D. (1959). Responses of cats to tactile and noxious stimuli. *Arch. Neurol. Psychiatry, Chicago*, 1: 203–215.

Halász, P. and Ujszászi, J. (1988). A study of K-complexes in humans: are they related to information processing during sleep? In: W.P. Koella, F. Obál, H.Shulz and P. Visser (Eds), *Sleep '86*. Gustav Fisher Verlag, Stuttgart/New York, pp. 79–83.

Hall, R.D. and Borbély, A.A. (1970). Acoustically evoked potentials in the rat during sleep and waking. *Exp. Brain Res.*, 11: 93–110.

Harsh, J., Voss, U., Hull, J., Schrepfer, S. and Badia, P. (1994). ERP and behavioral changes during the wake/sleep transition. *Psychophysiology*, 31: 244–252.

Held, H. (1897). Zur Kenntiss der peripheren Gehorleitung. *Arch. Anat. Physiol. Anat.*, pp. 350–360.

Hennevin, E., Huetz, C. and Edeline, J.-M. (2007). Neural representations during sleep: from sensory processing to memory traces. *Neurobiol. Learn. Mem.*, 87: 416–440.

Hernández-Peón, R., Chávez Ibarra, G., Morgane, J.P. and Timo Iaria, C. (1963). Limbic cholinergic pathways involved in sep and emotional behavior. *Exp. Neurol.*, 8: 93–111.

Herz, A. (1965). Cortical and subcortical auditory evoked potentials during wakefulness and sleep in cat. In: K. Akert, C. Bally and J.P. Shade (Eds), *Sleep Mechanisms. Progress in Brain Research*. Elsevier, Amsterdam, pp. 63–69.

Herz, A., Fraling, F., Niedner, I. and Farber, G. (1967). Pharmacologically induced alterations of cortical and subcortical evoked potentials compared with physiological changes during the awake–sleep cycle in cats. *Electroencephgr. Clin. Neurophysiol. Suppl.*, 26: 164–176.

Hess, W.R. (1944). Das schlafsyndrom als folge dienzephaler reizung. *Helv. Physiol. Pharamcol. Acta*, 2: 305–344.

Hobson, J.A. and McCarley, R.W. (1972). Spontaneous discharge rates of cat cerebellar Purkinje cells. *Electroencephgr. Clin. Neurophysiol.*, 33: 457–459.

Hobson, J.A., Pace-Schott, E.F., Stickgold, R. and Kahn, D. (1998). To dream or not to dream? Relevant data from new neuroimaging and electrophysiological studies. *Curr. Opin. Neurobiol.*, 8: 239–244.

Hu, B., Steriade, M. and Deschenes, M. (1989). The cellular mechanism of thalamic ponto-genicul-occipital waves. *Neurosci.*, 31: 25–35.

Huffman, R.F. and Henson, O.W. (1990). The descending auditory pathway and acousticomotor systems: connections with the inferior colliculus. *Brain Res. Rev.*, 15: 295–323.

Huttenlocher, P.R. (1960). Effects of the state of arousal on click responses in the mesencephalic reticular formation. *Electroencephgr. Clin. Neurophysiol.*, 12: 819–827.

Jeffress, L.A. (1948). A place theory of sound localization. *J. Comp. Physiol. Psychol.*, 41: 35–39.

John, E.R. (2001). The neurophysics of consciousness. *Brain Res. Rev.*, 39: 1–28.

John, E.R. (2006). The sometimes pernicious role of theory in science. *Int. J. Psychophysiol.*, 62: 377–383.

John, E.R. and Ranshoff, J. (1996). Coordination of international coma treatment consortium for NICHD.

Jones, B.E., Harper, S.T. and Halaris, A.E. (1977). Effects of locus coeruleus lesions upon cerebral monoamine content, sleep–wakefulness states and the response to amphetamine in the cats. *Brain Res.*, 124: 473–496.

Jouvet, M. (1961). Telencephalic and rombencephalic sleep in the cat. In: G.E.W. Wostenholme and C.M. O'Connor (Eds), *The Nature of Sleep*. Churchill, London.

Jouvet, M. (1962). Recherches sur les structures nerveuses et le mecanismes responsables de differentes phases du sommeil physiologique. *Arch. Ital. Biol.*, 100: 125–206.

Jouvet, M. (1999). *The Paradox of Sleep: The Story of Dreaming*. The MIT Press, Boston, MA.

Jouvet, M., Jeannerod, M. and Delorme, F. (1965). Organisation du systéme responsable de l'activité phasique au course du sommeil paradoxal. *C.R. Seances Soc. Biol. (Paris)*, 159: 1599–1604.

Kahana, M.J., Sekuler, R., Caplan, J.B., Kirschen, M. and Madsen, J.R. (1999). Human theta oscillations exhibit task dependence during virtual maze navigation. *Nature*, 399: 781–784.

Kahana, M.J., Seelig, D. and Madsen, J.R. (2001). Theta returns. *Curr. Opin. Neurobiol.*, 11: 739–744.

Kajimura, N., Uchiyama, M., Takayama, Y., Uchida, S., Uema, T., Kato, M., Sekimoto, M., Watanabe, T., Nakajima, T., Horikoshi, S., Ogawa, K., Nishikawa, M., Hiroki, M., Kudo, Y., Matsuda, H., Okawa, M. and Takahashi, K. (1999). Activity of midbrain reticular formation and neocortex during the progression of human non-rapid eye movement sleep. *J. Neurosci.*, 19: 10065–10073.

Kakigi, R., Naka, D., Okusa, T., Wang, X., Inui, K., Qiu, Y., Diep Tran, T., Miki, K., Tamura, Y., Nguyen, T.B., Watanabe, S. and Hoshiyama, M. (2003). Sensory perception during sleep in humans: a magnetoencephalograhic study. *Sleep Med.*, 4: 493–507.

Kasamatsu, T., Kiyono, S. and Iwama, K. (1967). Electrical activities of the visual cortex in chronically blinded cats. *Tohoku J. Exp. Med.*, 93: 139–152.

Kaufman, L.S. and Morrison, A.R. (1981). Spontaneous and elicited PGO spikes in rats. *Brain Res.*, 214: 61–72.

Kemp, D.T. (1978). Stimulated acoustic emissions from within the human auditory system. *J. Acoust. Soc. Am.*, 64: 1386–1391.

Khazan, N. and Sawyer, C.H. (1963). "Rebound" recovery from deprivation of paradoxical sleep in the rabbit. *Proc. Soc. Exp. Biol. Med.*, 114: 536–539.

Kirk, E. and Smith, D.W. (2003). Protection from acoustic trauma is not a primary function of the medial olivocochlear efferent system. *J. Assoc. Res. Otolaryngol.*, 4: 445–465.

Kleitman, N. (1963). *Sleep and Wakefulness.* University of Chicago Press, Chicago/London.

Klimesch, W. (1999). EEG alpha and theta oscillations reflect cognitive and memory performance: a review and analysis. *Brain Res. Rev.*, 29: 169–195.

Knudsen, E.I. and Konishi, M. (1978). A neural map of auditory space in the owl. *Science*, 4343: 795–797.

Kocsis, B. and Vertes, R.P. (1992). Dorsal raphe neurones: synchronous discharge with theta rhythm of the hippocampus in the freely behaving rat. *J. Neurophysiol.*, 68: 1463–1467.

Komisariuk, B. (1970). Synchrony between limbic system theta activity and rhythmical behaviour in rats. *J. Comp. Physiol. Psychol.*, 10: 482–492.

Kramis, R., Vanderwolf, C.H. and Bland, B.H. (1975). Two types of hippocampal rhythmical slow activity in both the rabbit and the rat: relations to behaviour and effects atropine, diethyl ether, urethane and pentobarbital. *Exp. Neurol.*, 49: 58–85.

Krueger, J.M. and Obal Jr., F. (1994). Sleep factors. In: N.A. Saunders and C.E. Sullivan (Eds), *Sleep and Breathing.* Marcel Dekker Inc., New York, pp. 79–112.

Kuhl, B.A., Dudukovic, N.M., Kahn, I. and Wagner, A.D. (2007). Decreased demands on cognitive control reveal the neural processing benefits of forgetting. *Nature Neurosci.*, 10: 908–914.

Lai, Y.Y. and Siegel, J.M. (1991). Pontomedullary glutamate receptors mediating locomotion and muscle tone suppression. *J. Neurosci.*, 11: 2931–2937.

Lee, A.K. and Wilson, M.A. (2002). Memory of sequential experience in the hippocampus during slow wave sleep. *Neuron*, 36: 1183–1194.

Lerma, J. and García-Austt, E. (1985). Hippocampal theta rhythm during paradoxical sleep. Effects of afferent stimuli and phase-relationships with phasic events. *Electroencephgr. Clin. Neurophysiol.*, 60: 46–54.

Levey, A.I., Hallanger, A.E. and Wainer, B.H. (1987). Choline acetyltransferase immunoreactivity in the rat thalamus. *J. Comp. Neurol.*, 257: 317–332.

Lewis, E.R. and Henry, K.R. (1992). Modulation of cochlear nerve spike rate by cardiac activity in the gerbil. *Hearing Res.*, 63: 7–11.

Liberman, M.C. (1990). Effects of chronic cochlear de-efferentation on auditory-nerve response. *Hear. Res.*, 49: 209–224.

Liberman, T., Velluti, R.A. and Pedemonte, M. (2006). Correlación entre neuronas auditivas del colículo inferior y el ritmo theta del hipocampo. *Physiol. Mini-Rev.*, 2: 65.

Libert, J.-P. and Bach, V. (2005). Thermoregulation and sleep in the human. In: P.L. Parmeggiani and R.A. Velluti (Eds), *The Physiologic Nature of Sleep*. Imperial College Press, London, pp. 407–431.

Livingstone, M.S. and Hubel, D.H. (1981). Effects of sleep and arousal on the processing of visual information in the cat. *Nature*, 291: 554–561.

Loewy, D.H., Campbell, K.B. and Bastien, C. (1996). The mismatch negativity to frequency deviant stimuli during natural sleep. *Electroencephgr. Clin. Neurophysiol.*, 98: 493–501.

Loomis, A.L., Harvey, E.N. and Hobart, G.A. (1938). Distribution of disturbance-patterns in the human electroencephalogram, with special reference to sleep. *J. Neurophysiol.*, 1: 413–430.

López, C., Rodríguez-Servetti, Z., Velluti, R.A. and Pedemonte, M. (2007). Influence of the olfactory system on the wakefulness–sleep cycle. *World Federation of Sleep Research Sleep Medicine Societies Congress (WFSRSMS)*, Cairns, Australia.

Lopes da Silva, F.H., Witter, M.P., Boeijinga, P.H., and Lohman, A.H.M. (1990). Anatomical organization and physiology of the limbic cortex. *Physiol. Rev.*, 70: 453–511.

Lorente de Nó, R. (1933). Anatomy of the eight nerve. III. General plan of structure of the primary cochlear nuclei. *Laryngoscope*, 43: 327–350.

Lorente de Nó, R. (1981). *The Primary Acoustic Nuclei*. Raven Press, New York.

Lorenzo, D., Velluti, J.C., Crispino, L. and Velluti, R.A. (1978). Cerebellar sensory functions: rat auditory evoked potentials. *Exp. Neurol.*, 55: 629–636.

Louie, K. and Wilson, M.A. (2001). Temporally structured replay of awake hippocampal ensemble activity during rapid eye movement sleep. *Neuron*, 29: 145–156.

Lu, J., Sherman, D., Devor, M. and Saper, C.B. (2006). A putative flip-flop switch for control of REM sleep. *Nature*, 441(1): 589–594.

Magee, J.C. (2003). A prominent role for intrinsic neuroneal properties in temporal coding. *Trends Neurosci.*, 26: 14–16.

Mano, N.I. (1970). Changes of simple and complex spike activity of cerebellar Purkinje cells with sleep and waking. *Science*, 170: 1325–1327.

Maquet, P. (2000). Functional neuroimaging of normal sleep by positron emission tomography. *J. Sleep Res.*, 9: 207–231.

Maquet, P., Dive, D., Salmon, E., Sadzot, B., Franco, G., Poirrier, R., von Frenckell, R. and Franck, G. (1990). Cerebral glucose utilization during sleep–wake cycle in man determined by positron emission tomography and [18F]2-fluoro-2-deoxy-D-glucose method. *Brain Res.*, 513: 136–143.

Maquet, P., Degueldre, C., Delfiore, G., Aerts, J., Peters, J.M., Luxen, A. and Franck, G. (1997). Functional neuroanatomy of human slow wave sleep. *J. Neurosci.*, 17: 2807–2812.

Maquet, P.A.A., Sterpenich, V., Albouy, G., Dang-vu, T., Desseilles, M., Boly, M., Huby, P., Laureys, S. and Peigneux, P. (2005). Brain imaging on passing to sleep. In: P.L. Parmeggiani and R.A. Velluti (Eds), *The Physiologic Nature of Sleep*. Imperial College Press, London, pp. 489–408.

Margrie, T.W. and Schaefer, A.T. (2003). Theta oscillation coupled spike latencies yield computational vigour in a mammalian sensory system. *J. Physiol.*, 546: 363–374.

Marks, G.A., Shaffery, J.P., Oksenberg, A., Speciale, S.G. and Roffwarg, H.P. (1995). A functional role for REM sleep in brain maturation. *Behav. Brain Res.*, 69: 1–11.

Martin, P. and Hudspeth, A.J. (1999). Active hair bundle movements can anplify a hair cells's response to oscillatory mechanical stimuli. *Proc. Acad. Sci. USA*, 96: 14306–14311.

Massaux, A.E. and Edeline, J.-M. (2003). Burst in the medial geniculate body: a comparison between anaesthetised and un anaesthetized guinea pig. *Exp. Brain Res.*, 153: 573–578.

McCarley, R.W. (2004). Mechanisms and models of REM sleep control. *Arch. Ital. Biol.*, 142: 429–467.

McCarley, R.W., Nelson, J.P. and Hobson, J.A. (1978). Ponto geniculo occipital (PGO) burst neurons: correlative evidence for neuronal generators of PGO waves. *Science*, 20: 269–272.

McCarley, R.W. and Hoffman, E.A. (1981). REM sleep dreams and the activation–synthesis hypothesis. *Am. J. Psychiatry*, 38: 904–912.

McCarley, R., Benoit, O. and Barrionuevo, G. (1983). Lateral geniculate nucleus unitary discharge in sleep and waking: state- and rate-specific aspects. *J. Neurophysiol.*, 50: 798–817.

McGinty, D. and Szymusiak, R. (2003). Hypothalamic regulation of sleepand arousal. *Front. Biosci.*, 960: 165–173.

McGinty, D., Alam, N., Suntsova, N., Guzman-Marin, R., Methippara, M., Gong, H. and Szymusiak, R. (2005). Hypothalamic mechanisms of sleep: perspective from neuronal unit recording studies. In: P.L. Parmeggiani and R.A. Velluti (Eds), *The Physiologic Nature of Sleep*. Imperial College Press, London, pp. 139–160.

McNaughton, N. and Morris, R.G. (1987). Chlordiazepoxide, an anxiolytic benzodiazepine, impairs place navigation in rats. *Behav. Brain Res.*, 24: 39–46.

Mehta, M.R., Lee, A.K. and Wilson, M.A. (2002). Role of experience and oscillations in transforming a rate code into a temporal code. *Nature*, 417: 741–746.

Mendel, M.I. and Goldstein, R. (1971). Early components of the averaged electro-encephalographic response to constant level clicks during all night sleep. *J. Speech Hear. Res.*, 14: 829–840.

Mignot, E. (2004). Sleep, sleep disorders and hypocretin (orexin). *Sleep Med.*, 5(Suppl. 1): S2–S8.

Mitani, A., Shimokouchi, M. and Nomura, S. (1983). Effects of the stimulation of the primary auditory cortex upon colliculo-geniculate neurons in the inferior colliculus of the cat. *Neurosci. Lett.*, 42: 185–189.

Moller, A.R. and Rollins, P.R. (2002). The non-classical auditory pathways are involved in hearing in children but not in adults. *Neurosci. Lett.*, 319: 41–44.

Monnier, M. and Hösli, L. (1964). Dialysis of sleep and waking factor in blood of the rabbit. *Science*, 146: 796–798.

Montero, V.M. (1983). Ultrastructure identification of axon terminals from thalamic reticular nucleus in the medial geniculate body in the rat: an EM autoradiographic study. *Exp. Brain Res.* 51: 338–342.

Morales-Cobas, G., Ferreira, M.I. and Velluti, R.A. (1995). Sleep and waking firing of inferior colliculus neurons in response to low frequency sound stimulation. *J. Sleep Res.*, 4: 242–251.

Morrison, A.R. and Pompeiano, O. (1966). Vestibular influences during sleep IV: Functional relations between vestibular nuclei and lateral geniculate nucleus during desynchronized sleep. *Arch Ital Biol.*, 104: 425–458.

Morrison, A.R. and Reiner, P.B. (1985). A dissection of paradoxical sleep. In: D. McGinty, R. Drucker-Colin, A.R. Morrison and P.L. Parmeggiani (Eds), *Brain Mechanisms of Sleep*. Raven Press, New York, pp. 97–110.

Moruzzi, G. (1960). Synchronizing influences of the brain stem and the inhibitory mechanisms underlying the production of sleep by sensory stimulation. *Electroencephgr. Clin. Neurophysiol. Suppl.*, 13: 231–256.

Moruzzi, G. (1963). Active processes in the brain stem during sleep. *Harvey Lect. Ser.*, 58: 253–297.

Moruzzi, G. (1972). The sleep–waking cycle. *Ergebnisse der Physiologie*, 64: 1–165.

Moruzzi, G. and Magoun, H. (1949). Brain stem reticular formation and activation of the EEG. *Electroencephgr. Clin. Neurophysiol.*, 1: 455–473.

Mountain, D.C. (1980). Changes in endolymphatic potential and crossed olivococlear bundle stimulation alter cochlear mechanics. *Science*, 210: 71–72.

Mukamel, R., Gelbard, H., Arieli, A., Hasson, U., Fried, I. and Malach, R. (2005). Coupling between neuronal firing, field potentials, and FMRI in human auditory cortex. *Science*, 5736: 951–954.

Mulders, W.H.A.M. and Robertson, D. (2000). Evidence for direct cortical innervation of medial olivocochlear neurons in rats. *Hearing Res.*, 144: 65–72.

Muzet, A. and Naitoh, P. (1977). Sommeil et bruit. *Confront. Psychiat.*, 15: 215–235.

Näätänen, R. (1992). *Attention and Brain Function*. Erlbaum, Hillsdale, NJ.

Näätänen, R. and Alho, K. (1995). Mismatch negativity – a unique measure of sensory processing in audition. *Int. J. Neurosci.*, 80: 317–337.

Naka, D., Kakigi, R., Hoshiyama, M., Yamasaki, H., Okusa, T. and Koyama, S. (1999). Structure of the auditory evoked magnetic fields during sleep. *Neuroscience*, 93: 573–583.

Narins, P.M. and Hurley, D.D. (1982). The relationship between call intensity and function in the Puerto Rican Coqui (Anura: Leptodactylidae). *Herpetologica*, 38: 287–295.

Neckelmann, D. and Ursin, R. (1993). Sleep stages and EEG power spectrum in relation to acoustical stimulus arousal threshold in the rat. *Sleep*, 16: 467–477.

Nick, T.A. and Konishi, M. (2001). Dynamic control of auditory activity during sleep: correlation between song response and EEG. *Proc. Natl. Acad. Sci. USA*, 98: 14012–14016.

Nielsen-Bohlman, L., Knight, R.T., Woods, D.L. and Woodward, K. (1991). Differential auditory processing continues during sleep. *Electroencephgr. Clin. Neurophysiol.*, 79: 281–290.

Nuñez, A., de Andrés, I. and García-Austt, E. (1991). Relationships of nucleus reticularis pontis oralis neuronal discharge with sensory and carbachol evoked hippocampal theta rhythm. *Exp. Brain Res.*, 87: 303–308.

O'Keefe, J. and Recce, M.L. (1993). Phase relationship between hippocampal place units and EEG theta rhythm. *Hippocampus*, 3: 317–330.

Oatman, L.C. (1971). Role of visual attention on auditory evoked potentials in unanesthetized cats. *Exp. Neurol.*, 32: 341–356.

Oatman, L.C. (1976). Effects of visual attention on the intensity of auditory evoked potentials. *Exp. Neurol.*, 51: 41–53.

Obal Jr., F. and Krueger, J.M. (2005). Humoral mechanisms of sleep. In: P.L. Parmeggiani and R.A. Velluti (Eds), *The Physiologic Nature of Sleep*. Imperial College Press, London, pp. 23–43.

Oliver, D.L. and Morest, D.K. (1984). The central nucleus of the inferior colliculus of the cat. *J. Comp. Neurol.*, 222: 237–264.

Oliver, D.L. and Shneiderman, A. (1991). The anatomy of the inferior colliculus: a cellular basis for integration of monaural and binaural information. In: R.A. Altshuler, R.P. Bobbin, B.M. Clopton and D.W. Hoffman (Eds), *Neurobiology of Hearing: Central Auditory System*. Raven Press, New York, pp. 195–222.

Ogilvie, R.D., Simons, I.A., Kuderian, R.H., MacDonald, T. and Rustenburg, J. (1991). Behavioral, event-related potential, and EEG/FFT changes at sleep onset. *Psychophysiology*, 28: 54–64.

Orem, J. (1989). Behavioral inspiratory inhibition: Inactivated and activated respiratory cells. *J. Neurophysiol.*, 62: 1069–1078.

Orem, J.M. (2005). Neural control of breathing in sleep. In: P.L. Parmeggiani and R.A. Velluti (Eds), *The Physiologic Nature of Sleep*. Imperial College Press, London.

Orem, J., Lovering, A.T., Dunin-Barkowski, W. and Vidruk, E.H. (2002). Tonic activity in the respiratory system in wakefulness, NREM and REM sleep. *Sleep*, 25: 488–496.

Ornitz, E.M., Panman, L.M. and Walter, R.D. (1967). The variability of the auditory averaged evoked responses during sleep and dreaming in children and adults. *Electroencephgr. Clin. Neurophysiol.*, 22: 514–524.

Osterhammel, P., Shallop, J. and Terkildsen, K. (1985). The effects of sleep on the auditory brainstem response (ABR) and the middle latency response (MLR). *Scand. Audiol.*, 14: 47–50.

Oswald, I. (1960). Falling asleep open eyed during intense rhythmic stimulation. *Br. Med. J.*, 1: 1450–1455.

Oswald, I., Taylor, A.M. and Treisman, M. (1960). Discriminative responses to stimulation during human sleep. *Brain*, 83: 440–453.

Parmeggiani, P.L. (1980). Temperature regulation during sleep: a study in homeostasis. In: H Orem and Ch. Barnes (Eds), *Physiology in Sleep*, Academic Press, New York, pp. 98–136.

Parmeggiani, P.L. (2005a). Sleep behaviour and temperature. In: P.L. Parmeggiani and R.A. Velluti (Eds), *The Physiologic Nature of Sleep*. Imperial College Press, London, pp. 387–406.

Parmeggiani, P.L. (2005b). Physiologic regulation in sleep. In: M.H. Kryger, T. Roth and W.C. Dement (Eds), *Principles and Practice of Sleep Medicine*. Elsevier Saunders, Philadelphia, pp. 185–191.

Parmeggiani, P.L. and Rapisarda, C. (1969). Hippocampal output and sensory mechanisms. *Brain Res.*, 14: 387–400.

Parmeggiani, P.L. and Sabattini, L. (1972). Electromyographic aspects of postural, respiratory and thermoregulatory mechanisms in sleeping cats. *Electroencephgr. Clin. Neurophysiol.*, 33: 1–13.

Parmeggiani, P.L., Lenzi, P., Azzaroni, A. and D'Alessandro, R. (1982). Hippocampal influence on unit responses elicited in the cat's auditory cortex by acoustic stimulation. *Exp. Neurol.*, 78: 259–274.

Pavez, E., Drexler, D., Délano, P.H., Pedemonte, M., Falconi, A., Robles, L. and Velluti, R.A. (2006). Efectos del ruido sobre potenciales cocleares en chinchilla y cobayo. *Physiological Mini-Rev.*, 2: 64.

Pearson, K.S., Berber, D.S., Tabachnick, B.G. and Fidell, S. (1995). Predicting noise-induced sleep disturbances. *J. Acoust. Soc. Am.*, 97: 331–338.

Pedemonte, M. and Velluti, R.A. (2005a). What individual neurons tell us about encoding and sensory processing in sleep. In: P.L. Parmeggiani and R.A. Velluti (Eds), *The Physiologic Nature of Sleep*. Imperial College Press, London, pp. 489–508.

Pedemonte, M. and Velluti, R.A. (2005b). Sleep hippocampal theta rhythm and sensory processing. In: M. Lander, D.P. Cardinali and P. Perumal (Eds), *Sleep and Sleep Disorders: A Neuropsychopharmacological Approach*. Landes Biosciencies, TX/Springer, NY, pp. 8–12.

Pedemonte, M., Peña, J.L. and Velluti, R. (1990). Periaqueductal gray influence on anteroventral cochlear nucleus unitary activity and naloxone effects. *Hear. Res.*, 47: 219–228.

Pedemonte, M., Peña, J.L., Morales-Cobas, G. and Velluti, R.A. (1994). Effects of sleep on the responses of single cells in the lateral superior olive. *Arch. Ital. Biol.*, 132: 165–178.

Pedemonte, M., Peña, J.L., Bouvier, M. and Velluti, R.A. (1995). Differential modulation of auditory signals by hippocampal theta rhythm in sleep phases and waking. *Sleep Res.*, 24A: 52.

Pedemonte, M., Peña, J.L. and Velluti, R.A. (1996a). Firing of inferior colliculus auditory neurone is phase-locked to the hippocampus theta rhythm during paradoxical sleep and waking. *Exp. Brain Res.*, 112: 41–46.

Pedemonte, M., Peña, J.L., Torterolo, P. and Velluti, R.A. (1996b). Auditory deprivation modifies sleep in the guinea-pig. *Neurosci. Lett.* 223: 1–4.

Pedemonte, M., Torterolo, P. and Velluti, R.A. (1997). *In vivo* intracellular characteristics of inferior colliculus neurons in guinea pigs. *Brain Res.*, 759: 24–31.

Pedemonte, M., Rodríguez, A. and Velluti, R.A. (1999). Hippocampal theta waves as an electrocardiogram rhythm timer in paradoxical sleep. *Neurosci. Lett.*, 276: 5–8.

Pedemonte, M., Pérez-Perera, L., Peña, J.L. and Velluti, R.A. (2001). Sleep and wakefulness auditory processing: cortical units vs. hippocampal theta rhythm. *Sleep Res. Online*, 4: 51–57.

Pedemonte, M., Goldstein-Daruech, N. and Velluti, R.A. (2003). Temporal correlation between heart rate, medullary units and hippocampal theta rhythm in anesthetized, sleeping and awake guinea pigs. *Auton. Neurosci. Basic Clin.*, 107: 99–104.

Pedemonte, M., Drexler, D.G. and Velluti, R.A. (2004). Cochlear microphonic changes after noise exposure and gentamicin administration during sleep and waking. *Hear. Res.*, 194: 25–30.

Pedemonte, M., Gambini, J.P. and Velluti, R.A. (2005). Novelty-induced correlation between visual neurons and the hippocampal theta rhythm in sleep and wakefulness. *Brain Res.*, 1062: 9–15.

Pfeiffer, R.R. (1966). Anteroventral cochlear nucleus: waveforms of extracellularly recorded spike potentials. *Science*, 154: 667–668.

Peña, J.L. and Konishi, M. (2001). Auditory spatial receptive fields created by multiplication. *Science*, 292: 249–252.

Peña, J.L., Pedemonte, M., Ribeiro, M.F. and Velluti, R. (1992). Single unit activity in the guinea-pig cochlear nucleus during sleep and wakefulness. *Arch. Ital. Biol.*, 130: 179–189.

Peña, J.L., Pérez-Perera, L., Bouvier, M. and Velluti, R.A. (1999). Sleep and wakefulness modulation of the neuronal firing in the auditory cortex of the guinea-pig. *Brain Res.*, 816: 463–470.

Pérez-Perera, L. (2002). Actividad unitaria de la corteza auditiva: ritmo theta del hipocampo y respuesta a vocalizaciones en el ciclo vigilia-sueño. *Master Thesis*, Montevideo.

Pérez-Perera, L., Bentancor, C., Pedemonte, M. and Velluti, R.A. (2001). Auditory cortex unitary activity correlated to sleep–wakefulness and theta rhythm in response to natural sounds. *Actas de Fisiología*, 7: 187.

Peyret, D., Campistron, G., Geffard, M. and Aran, J.M. (1987). Glycine immunoreactivity in the brainstem auditory and vestibular nucluei of the guinea pig. *Acta Otolaryngol. (Stockh)*, 104: 71–76.

Perrin, F., García-Larrea, L., Mauguière, F. and Bastuji, H. (1999). A differential brain response to the subject's own name persist during sleep. *Clin. Neurophysiol.*, 110: 2153–2164.

Petrek, J., Golda, V. and Lisonek, P. (1968). Cortical response amplitude changes produced by rhythmic acoustic stimulation in cats. *Exp. Brain Res.*, 6: 19–31.

Picton, T.W., Hillyard, S.A., Krausz, H.I. and Galambos, R. (1974). Human auditory evoked potentials. I. Evaluation of components. *Electroenceph. Clin. Neurophysiol.*, 36: 179–190.

Pieron, H. (1913). *Le Problème Physiologique du Sommeil*. Masson & Cie., Paris.

Pompeiano, O. (1967). The neurophysiological mechanisms of the postural and motor events during desynchronized sleep. *Proc. Assoc. Res. Nerve. Ment. Dis.*, 45: 351–423.

Pompeiano, O. (1970). Mechanisms of sensorimotor integration during sleep. In: E. Stellar and J.M. Sprague (Eds), *Progress in Physiological Psychology*. Academic Press, New York/London, pp. 1–179.

Porkka-Heiskanen, T., Strecker, R.E., Thakkar, M., Bjorkum, A.A., Greene, R.W. and McCarley, R.W. (1997). Adenosine: a mediator of the sleep-inducing effects of prolonged wakefulness. *Science*, 276: 1265–1268.

Portas, Ch.M. (2005). Cognitive aspects of sleep: perception, mentation, and dreaming. In: P.L. Parmeggiani and R.A. Velluti (Eds), *The Physiologic Nature of Sleep*. Imperial College Press, London, pp. 535–569.

Portas, C.M., Krakow, K., Allen, P., Josephs, O., Armony, J.L. and Frith, C.D. (2000). Auditory processing across the sleep–wake cycle: simultaneous EEG and fMRI monitoring in human. *Neuron*, 28: 991–999.

Pradham, S.N. and Galambos, R. (1963). Some effects of anesthetics on the evoked responses in the auditory cortex of cats. *J. Pharmacol. Exp. Ther.*, 139: 97–106.

Rabat, A., Bouyer, J.J., Aran, J.M., Courtiere, A., Mayo, W. and LeMoal, M. (2004). Deleterious effects of an environmental noise on sleep and contribution of its physical components in a rat model. *Brain Res.*, 1009: 88–97.

Radmilovich, M., Bertolotto, C., Peña, J.L., Pedemonte, M. and Velluti, R.A. (1991). A search for e mesencephalic pariaqueductal gray-cochlear nucleus connection. *Acta Physiol. Pharmacol. Latinoam.*, 41: 369–375.

Ramón y Cajal, S. (1952). *Histologie du Système Nerveux*. Consejo Superior de Investigaciones Científicas. Madrid, España.

Rasmussen, G.L. (1946). The olivary peduncle and other fiber projections of the superior olivary complex. *J. Comp. Neurol.*, 84: 141–219.

Rasmussen, G.L. (1960). Efferent fibers of cochlear nerve and cochlear nucleus. In: G.L. Rasmussen and W.F. Windle (Eds), *Neural Mechanisms of the Auditory and Vestibular Systems*. Thomas, Springfield, IL, pp. 105–115.

Rawlins, J.N.P. (1985). Associations across time in the hippocampus as a temporary memory store. *Behav. Brain Sci.*, 8: 479–496.

Recanzone, G. (2000). Response profiles of auditory cortical neurons to tones and noise in behaving macaque monkeys. *Hear. Res.*, 150: 104–118.

Redding, F.K. (1967). Modification of sensory cortical evoked potentials by hippocampal stimulation. *Electroencephgr. Clin. Neurophysiol.*, 22: 74–83.

Reiner, P.B. and Morrison, A.R. (1980). Pontine-geniculate-occipital spikes in the albino rat: Evidence for the presence of the pontine component as revealed by cerebellar lesions. *Exp Neurol.*, 69: 61–73.

Reinoso-Suárez, F., De Andrés, I., Rodrigo-Angulo, M.L. and Rodríguez-Veiga, E. (1994). Location and anatomical connections of a paradoxical sleep induction site in the cat ventral pontine tegmentum. *Eur. J. Neurosci.*, 6: 1829–1836.

Reivich, M. (1974). Blood flow metabolism couple in brain. In: F. Plum (Ed.), *Brain Dysfunction in Metabolic Disorders*. Raven, New York, pp. 125–140.

Rieke, F., Warland, D., de Ruyter, R. and Bialek, W. (1997). *Spikes. Exploring the Neural Coding*. Massachsetts Institute of Technology, USA.

Robles, L. and Ruggero, M.A. (2001). Mechanics of the mammalian cochlea. *Physiol. Rev.*, 81: 1305–1352.

Roffwarg, H.P., Muzio, J.N. and Dement, W.C. (1966). Ontogenetic development of the human sleep-dream cycle. *Science*, 152: 604–619.

Saberi, K., Takahashi, Y., Konishi, M., Albeck, Y., Arthur, B.J. and Farahbod, H. (1998). Effects of interaural decorrelation on neural and behavioral detection of spatial cues. *Neuron*, 21: 789–798.

Sakurai, Y. (1999). How do cell assemblies encode information in the brain? *Neurosci. Biobehav. Rev.*, 23: 785–796.

Saldaña, E. and Merchán, M.A. (1992). Intrinsic and commissural connections of the rat inferior colliculus. *J. Comp. Neurol.*, 319: 417–437.

Saldaña, E., Feliciano, M. and Mugnaini, E. (1996). Distribution of descending projections from primary auditory neocortex to inferior colliculus mimics the topography of intracollicular projections. *J. Comp. Neurol.*, 371: 15–40.

Sante de Sanctis (1899). *I sogni*. Fratelli Bocca, Torino.

Saper, C.B., Scamelli, T.E. and Lu, J. (2005). Hypothalamic regulation of sleep and circadian rhythms. *Nature*, 437: 1257–1263.

Sassin, J.F., Parker, D.C., Mace, J.W., Grotlin, R.W., Johnson, L.C. and Rossman, L.G. (1969). Human growth hormone release : relation to slow-wave sleep and sleep–waking cycles. *Science*, 165: 513–515.

Sauerland, E., Velluti, R. and Harper, R. (1972). Cortically induced changes of presynaptic excitability in higher-order auditory afferents. *Exp. Neurol.*, 36: 79–87.

Segundo, J.P., Moore, G.P., Stensaas, L.J. and Bullock, T.H. (1963). Sensitivity of neurons in *Aplysia* to temporal pattern of arriving impulses. *J. Exp. Biol.*, 40: 643–667.

Schofield, B.R. and Coomes, D.L. (2005). Auditory cortical projections to the cochlear nucleus in guinea pigs. *Hear. Res.*, 199: 89–102.

Schreiner, C.E., Read, H.L. and Sutter, M.L. (2000). Modular organization of frequency integration in primary auditory cortex. *Annu. Rev. Neurosci.*, 23: 501–509.

Semple, M.N. and Aitkin, L.M. (1979). Representation of sound frequency and laterality by units in central nucleus of cat inferior colliculus. *J. Neurophysiol.*, 42: 1626–1639.

Semple, M.N. and Scott, B.H. (2003). Cortical mechanisms in hearing. *Curr. Opin. Neurobiol.*, 13: 167–173.

Shannon, C.E. (1948). A mathematical theory of communication. *Bell Sys. Tech. J.*, 27: 379–423.

Shouse, M.N. and Siegel, J.M. (1992). Proteine regulation of REM sleep components in cats: integrity of the pedunculopontine tegmentum (PPT) is important for phasic events but unnecessary for atonia during REM sleep. *Brain Res.*, 571: 50–63.

Siegel, J.H. and King, D.O. (1982). Efferent neural control of cochlear mechanics? Olivococlear bundle stimulation affects cochlear biomechanical nonlinearity. *Hear. Res.*, 6: 171–182.

Silvani, A. and Lenzi, P. (2005). Reflex cardiovascular control in sleep. In: P.L. Parmeggiani and R.A. Velluti (Eds), *The Physiologic Nature of Sleep*. Imperial College Press, London, pp. 323–349.

Skaggs, W.E., McNaughton, B.L., Wilson, M.A. and Barnes, C. (1996). Theta phase precession in hippocampal neuroneal populations and the compression of temporal sequences. *Hippocampus*, 6: 149–172.

Soca, F. (1900). Sur un cas de sommeil prolongé pendant sept mois par un tumeur de l'hypophyse. *Nouv. Iconogr. Salpêtriére*, 2: 101–115.

Soja, P.J., Cairns, B.E. and Kristensen, M.P. (1998). Transmission through ascending trigeminal and lumbar sensory pathways: dependence on behavioral state. In: R. Lydic and H.A. Baghdoyan (Eds), *Handbook of Behavioral State Control*. CRC Press, Boca Raton/London/New York/Washington, pp. 521–544.

Spangler, K.M. and Warr, W.B. (1991). The descending auditory system. In: R.A. Altshuler, R.P. Bobbin and B.M. Clopton (Eds), *Neurobiology of Hearing. The Central Auditory System*. Plenum Press, New York, pp. 27–45.

Spitzer, M. (1999). The Brain Within the Net. MIT Press, Cambridge, MA.

Steriade, M. (2005). Brain electrical activity and sensory processing during waking and sleep states. In: M.H. Kryger, T. Roth and W.C. Dement (Eds), *Principles and Practice of Sleep Medicine*. Saunders Company, Philadelphia, PA.

Stickgold, R. (1998). Sleep of-line memory reprocessing. *Trends Cogn. Sci.*, 2: 484–492.

Stickgold, R., James, L. and Hobson, J.A. (2000a). Visual discrimination learning requires sleep after training. *Nat. Neurosci.*, 3: 1237–1238.

Stickgold, R., Malia, A., Maguire, D., Roddenberry, D. and O'Connor, M. (2000b). Replaying the game: hypnagogic images in normals and amnesics. *Science*, 290: 350–353.

Sun, X., Jen, P.H.S., Sun, D. and Zhang, G. (1989). Corticofugal influences on the responses of bat inferior collicular neurons to sound stimulation. *Brain Res.*, 495: 1–8.

Suntsova, N., Szymusiak, R., Alam, Md.N., Guzmán-Marín, R. and McGinty, D. (2002). Sleep–waking discharge patterns of medial preoptic nucleus neurons in rats. *J. Physiol.*, 543: 665–677.

Syka, J. and Popelar, J. (1984). Inferior colliculus in the rat: neuronal responses to stimulation of the auditory cortex. *Neurosci. Lett.*, 31: 235–240.

Tafti, M. (2003). Deficiency in short chain fatty acid beta-oxidation affects theta oscillations during sleep. *Nat. Genet.*, 34: 320–325.

Taheri, S., Zeitzer, J.M. and Mignot, E. (2002). The role of hypocretins (orexins) in sleep regulation and narcolepsy. *Annu. Rev. Neurosci.*, 25: 283–313.

Takahashi, Y., Kipnis, D.M. and Daughaday, W.H. (1968). Growth hormone secretion during sleep. *J. Clin. Invest.*, 47: 2079–2090.

Tammer, R., Ehrenreich, L. and Jürgens, U. (2004). Telemetrically recorded neuronal activity in the inferior colliculus and bordering tegmentum during vocal communication in squirrel monkeys (*Saimiri sciureus*). *Behav. Brain Res.*, 151: 331–336.

Tanaka, H., Fujita, N., Takanashi, M., Hirabuki, N., Yoshimura, H., Abe, K. and Nakamura, H. (2003). Effect of stage 1 sleep on auditory cortex during pure tone stimulation: evaluation by functional magnetic resonance imaging with simultaneous EEG monitoring. *AJNR Am. J. Neuroradiol.* 24: 1982–1988.

Teas, D.C. and Kiang, N.Y.S. (1964). Evoked responses from the auditory cortex. *Exp. Neurol.*, 10: 91–119.

Terzano, M.G., Parrino, L., Fioriti, G., Orofiamma, B. and Depoortere, H. (1990). Modifications of sleep structure by increasing levels of acoustic perturbation in normal subjects. *Electroenceph. Clin. Neurophysiol.*, 76: 29–38.

Tononi, G. and Cirelli, C. (2003). Sleep and synaptic homeostasis: a hypothesis. *Brain Res. Bull.*, 62: 143–150.

Tononi, G. and Cirelli, C. (2005). A possible role for sleep in synaptic homeostasis. In: P.L. Parmeggiani, R.A. Velluti (Eds). *The Physiologic Nature of Sleep*. Imperial College Press, London, pp. 77–101.

Torterolo, P., Pedemonte, M. and Velluti, R.A. (1995). Intracellular *in vivo* recording of inferior colliculus auditory neurons from awake guinea-pigs. *Arch. Ital. Biol.*, 134: 57–64.

Torterolo, P., Zurita, P., Pedemonte, M. and Velluti, R.A. (1998). Auditory cortical efferent actions upon inferior colliculus unitary activity. *Neurosci. Lett.*, 249: 172–176.

Ujszászi, J. and Halász, P. (1986). Late component variants of single auditory evoked responses during NREM sleep stage 2 in man. *Electroencephgr. Clin. Neurophysiol.*, 64: 260–268.

Vallet, M. (1982). La perturbation du sommeil par le bruit. *Soz. Praventivmed.*, 27: 124–131.

Vallet, M. and Mouret, J. (1984). Sleep disturbance due to transportation noise: ear plugs vs oral drugs. *Experientia*, 40: 429–437.

Van Twyvert, H.B., Levitt, R.A. and Dunn, R.S. (1966). The effect of high intensity white noise on the sleep pattern of the rat. *Psychon. Sci.*, 6: 355–356.

Vanzulli, A., Bogacz, J., García-Austt, E. (1961). Evoked responses in man. III. Auditory response. *Acta Neurol. Latinoam.*, 7: 303–308.

Velluti, R.A. (1985). An electrochemical approach to sleep metabolism: a pO2 paradoxical sleep system. *Physiol. Behav.*, 34: 355–358.

Velluti, R. (1988). A functional viewpoint on paradoxical sleep-related brain regions. *Acta Physiol. Pharmacol. Latinoam.*, 38: 99–115.

Velluti, R.A. (1997). Interactions between sleep and sensory physiology. A review. *J. Sleep Res.*, 6: 61–77.

Velluti, R.A. (2005). Remarks on sensory neurophysiological mechanisms participating in active sleep processes. In: P.L. Parmeggiani and R.A. Velluti (Eds), *The Physiologic Nature of Sleep*. Imperial College Press, London, pp. 247–265.

Velluti, R.A. and Crispino, L. (1979). Cerebellar actions on cochlear microphonics and on auditory nerve action potential. *Brain Res. Bull.*, 4: 621–624.

Velluti, R. and Galambos, R. (1970). Evoked resistance shifts: acute and chronic experiments at cortical and subcortical structures. *Electroencephgr. Clin. Neurophysiol.*, 28: 519.

Velluti, R. and Hernández-Peón, R. (1963). Atropine blockade within a cholinergic hypnogenic circuit. *Exp. Neurol.*, 8: 20–29.

Velluti, R. and Monti, J.M. (1976). pO$_2$ recorded in the amygdaloid complex during the sleep–waking cycle in cats. *Exp. Neurol.*, 50: 798–805.

Velluti, R.A. and Pedemonte, M. (1986). Differential effects of benzodiazepines on choclear and auditory nerve responses. *Electroencephgr. Clin. Neurophysiol.*, 64: 556–562.

Velluti, R.A. and Pedemonte, M. (2002). *In vivo* approach to the cellular mechanisms for sensory processing in sleep and wakefulness. *Cell Mol. Neurobiol.*, 22: 501–516.

Velluti, R.A. and Pedemonte, M. (2006). Sleep and sensory information. *Physiol. Mini-Rev.*, 2: 1–10.

Velluti, R., Roig, J.A., Escarcena, L.A., Villar, J.I. and García Austt, E. (1965). Changes of brain pO$_2$ during arousal and aletness in unrestrained cats. *Acta Neurol. Latinoam.*, 11: 368–382.

Velluti, R., Velluti, J.C. and García-Austt, E. (1977). Cerebellum pO$_2$ and the sleep–waking cycle in cats. *Physiol. Behav.*, 18: 19–23.

Velluti, R.A., Yamuy, J., Hadjez, J. and Monti, J.M. (1985). Spontaneous cerebellar nuclei PGO-like waves in natural paradoxical sleep and under reserpine. *Electroencephgr. Clin. Neurophysiol.* 60: 243–248.

Velluti, R.A., Pedemonte, M. and García-Austt, E. (1989). Correlative changes of auditory nerve and microphonic potentials throughout sleep. *Hear. Res.*, 39: 203–208.

Velluti, R.A., Pedemonte, M. and Peña, J.L. (1990). Auditory brain stem unit activity during sleep phases. In: J. Horne (Ed.), *Sleep '90*. Pontenagel Press, Bochum, pp. 94–96.

Velluti, R.A., Pena, J.L., Pedemonte, M. and Narins, P.M. (1994). Internally-generated sound stimulate cochlear nucleus units. *Hear. Res.*, 72: 19–22.

Velluti, R.A., Pedemonte, M. and Peña, J.L. (2000). Reciprocal actions between sensory signals and sleep. *Biol. Signals Recept.*, 9: 297–308.

Velluti, R.A., Pedemonte, M., Suárez, H., Inderkum, A., Rodríguez-Servetti, Z. and Rodríguez-Alvez, A. (2003). Human sleep architecture shifts due to auditory sensory input. *Sleep*, 26(Suppl.): A19.

Vertes, R.P. and Kocsis, B. (1997). Brainstem–diencephalo–septohippocampal systems controlling the theta rhythm of the hippocampus. *Neuroscience*, 81: 893–926.

Vinogradova, O.S. (2001). Hippocampus as comparator: role of the two input and two output systems of the hippocampus in selection and registration of information. *Hippocampus*, 11: 578–598.

Vital-Durand, F. and Michel, F. (1971). Effets de la desafferentation periphérique sur le cicle veille-sommeil chez le chat. *Arch. Ital. Biol.*, 109: 166–186.

Vivaldi, E., McCarley, R.W. and Hobson, J.A. (1980). Evocation of desynchronized sleep signs by chemical microstimulation of the pontine brain stem. In: J.A. Hobson and M.A.B. Brazier (Eds), *The Reticular Formation Revisited*. Raven Press, New York, pp. 513–529.

von Economo, C. (1930). Sleep as a problem of localization. *J. Nerv. Ment. Dis.*, 7: 249–259.

Wallenstein, G.W., Eichenbaum, H. and Hasselmo, M.E. (1998). The hippocampus as an associator of discontiguous events. *Trends Neurosci.*, 21: 317–323.

Warr, W.B. (1975). Olivocochlear and vestibular efferent neurons of the feline brain stem. Their location, morphology, and number determined by retrograde axonal transport and acetylcholinesterase histochemistry. *J. Comp. Neurol.*, 161: 159–182.

Warr, W.B. (1992). Organization o f olivococlear efferent system in mammals. In: D.B. Webster, A.N. Popper and R.R. Fay (Eds), *Mammalian Auditory Pathway: Neuroanatomy*. Springer Verlag, New York, pp. 410–448.

Weitzman, E.D. and Kremen, H. (1965). Auditory evoked responses during different stages of sleep in man. *Electroencephgr. Clin. Neurophysiol.*, 18: 65–70.

Wenthold, R.J. and Martin, M.R. (1984). Neurotransmitters of the auditory nerve and central auditory system. In: C. Berlin (Ed.). *Hearing Science: Recent Advances*. College-Hill Press, San Diego, CA, pp. 341–369.

Wesensten, N.J. and Badia, P. (1988). The P300 component in sleep. *Physiol. Behav.*, 44: 215–220.

Wickelgren, W.O. (1968). Effects of the state of arousal on click-evoked responses in cats. *J. Neurophysiol.*, 31: 757–768.

Williams, H.L., Tepas, D.I. and Morloch, H.C. (1963). Evoked responses to clicks and electroencephalographic stages of sleep in man. *Science*, 138: 685–686.

Winer, J.A. (1991). Anatomy of the medial geniculate body. In: R.A. Altshuler, R.P. Bobbin, B.M. Clopton and D.W. Hoffman (Eds), *Neurobiology of Hearing: Central Auditory System*. Raven Press, New York, pp. 293–333.

Winer, J.A. (2006). Decoding the auditory corticofugal systems. *Hear. Res.*, 212: 1–8.

Winer, J.A., Larue, D.T., Diehl, J.J. and Hefti, B.J. (1998). Auditory cortical projections to the cat inferior colliculus. *J. Comp. Neurol.*, 400: 147–174.

Winer, J.A., Diehl, J.J. and Larue, D.T. (2001). Projections of auditory cortex to the medial geniculate body of the cat. *J. Comp. Neurol.*, 430: 27–55.

Winson, J. (1978). Loss of hippocampal theta rhythm results in spatial memory deficit in the rat. *Science*, 201: 160–163.

Yamamoto, T., Lyeth, B.G., Dixon, E.D., Robinson, S.E., Jenkins, L.W., Young, H.E., Stonnington, H.H. and Hayes, R.L. (1988). Changes in regional brain acetylcholine content in rats following unilateral and bilateral brainstem lesions. *J. Neurotrauma*, 5: 69–79.

Zatorre, R.J., Bouffard, M., Ahad, P. and Belin, P. (2002). Where is "where" in the human auditory cortex? *Nat. Neurosci.*, 5: 906–909.

Zoccoli, G., Bojic, T. and Franzini, C. (2005). Regulation of cerebral circulation during sleep. In: P.L. Parmeggiani and R.A. Velluti (Eds), *The Physiologic Nature of Sleep*. Imperial College Press, London, pp. 351–369.

Zubek, J.P. (1969). Physiological and biochemical effects. In: J.P. Zubek (Ed.), *Sensory Deprivation: Fifteen Years of Research*. Appleton-Century-Crofts, New York, pp. 255–288.

Index

Printed and bound by CPI Group (UK) Ltd, Croydon, CR0 4YY

03/10/2024

01040415-0002